Applied Survival Analysis

WILEY SERIES IN PROBABILITY AND STATISTICS

Established by WALTER A. SHEWHART and SAMUEL S. WILKS

Editors: *Vic Barnett, Ralph A. Bradley, Noel A. C. Cressie, Nicholas I. Fisher, Iain M. Johnstone, J. B. Kadane, David G. Kendall, David W. Scott, Bernard W. Silverman, Adrian F. M. Smith, Jozef L. Teugels, Geoffrey S. Watson; J. Stuart Hunter, Emeritus*

A complete list of the titles in this series appears at the end of this volume.

Applied Survival Analysis

CHAP T. LE
Professor of Biostatistics, School of Public Health
Director of Biostatistics, Cancer Center
University of Minnesota

A Wiley-Interscience Publication
JOHN WILEY & SONS, INC.
New York · Chichester · Weinheim · Brisbane · Singapore · Toronto

This book is printed on acid-free paper. ∞

Copyright ©1997 by John Wiley & Sons, Inc. All rights reserved.

Published simultaneously in Canada.

No part of this publication may be reproduced, stored in a retrieval system or transmitted in any form or by any means, electronic, mechanical, photocopying, recording, scanning or otherwise, except as permitted under Sections 107 or 108 of the 1976 United States Copyright Act, without either the prior written permission of the Publisher, or authorization through payment of the appropriate per-copy fee to the Copyright Clearance Center, 222 Rosewood Drive, Danvers, MA 01923, (508) 750-8400, fax (508) 750-4744. Requests to the Publisher for permission should be addressed to the Permissions Department, John Wiley & Sons, Inc., 605 Third Avenue, New York, NY 10158-0012, (212) 850-6011, fax (212) 850-6008, E-Mail: PERMREQ@WILEY.COM.

Library of Congress Cataloging-in-Publication Data:

Le, Chap T., 1948–
   Applied survival analysis/Chap T. Le.
     p. cm. – (Wiley series in probability and mathematical statistics. Applied probability and statistics)
   "A Wiley-Interscience publication."
   Includes bibliographical references and index.
   ISBN 0-471-17085-2 (paper: alk. paper)
   1. Survival analysis (Biometry). I. Title. II. Series.
R853.S7L4 1997
610'.72–dc21

97-6506
CIP

Printed in the United States of America

10 9 8 7 6 5 4 3 2

To
MinhHa, Mina, and Jenna

Contents

1. Basic Concepts in Survival Analysis 1
    1.1. Study Designs, 2
    1.2. Survival Time, 3
    1.3. Functions of Survival Time, 4
    1.4. Regression Analysis, 12
    1.5. Examples of Survival Data, 13
    1.6. Examples of Survival Analysis, 16
            Empirical Cumulative Distribution Function, 17
            Comparison of Two Treatments, 20
            Comparison of Several Treatments, 23
            Rank Correlations, 28
            $2 \times k$ Contingency Tables, 31
            Maximum-Likelihood Estimation (MLE), 33
    1.7. Special Problem: Incomplete Data, 37
    1.8. Data Presentation: Event Chart, 39
    1.9. Number of Deaths, 41
    1.10. Computational Implementation, 43
   Exercises, 44

2. Estimation of Functions and Parameters 51
    2.1. Estimation of the Survival Function, 52
            Kaplan–Meier Curve, 53
            Actuarial Method, 57
            Mean, Median, and Other Parameters, 60
            Standardization of Survival Rates, 62
            Current Population Life Tables, 66

2.2. Estimation of Parameters in Survival Models, 67
  Some Survival Models, 68
  Likelihood Function, 69
  Exponential Model, 70
  Weibull Model, 72
  Special Use of the Standardized Mortality Ratio, 74
2.3. Cohort Studies, 77
  Method, 77
  Model, 80
  Results, 80
  Healthy Worker Effects, 84
2.4. Estimation of the Cumulative Hazard, 86
2.5. Rank Correlation for Bivariate Censored Data, 87
Exercises, 90

3. Comparison of Survival Distributions 97

3.1. Nonparametric Methods, 98
  Comparison of Two Groups, 99
  Comparison of Several Groups, 107
  Detection of Crossing-Curves Alternatives, 109
  Analysis of Pair-Matched Data, 112
3.2. Comparison of Two Exponential Distributions, 114
  Score Test, 115
  Cox F Test, 118
  Sample Size Determination, 121
  Methods for Person-Years Data, 125
3.3. Piecewise Exponential: Link Between Parametric and Nonparametric Methods, 128
3.4. Trends in Survival, 132
  Class of Tests Against Stochastically Ordered Alternatives, 133
  Tarone Test, 137
  Le–Gramsch–Louis Method, 138
  Test for Trend in Constant Hazards and Its Application in Occupational Health, 142
3.5. Comparison of Two Clustered Samples, 148
  Problem, 148
  Data Summarization, 149

Models, 151
Test Statistic, 153
Exercises, 155

4. Correlation and Regression Analyses 161

4.1. Simple Regression and Correlation, 163
Model and Approach, 163
Measures of Association, 165
Effects of Measurement Scale, 167
Tests of Association, 169

4.2. Multiple Regression, 176
Proportional Hazards Models with Several Covariates, 177
Testing Hypotheses in Multiple Regression, 182
Stepwise Procedure, 187
Estimation of the Baseline Survival Function, 192

4.3. Time-Dependent Covariates, 194
Examples, 195
Implementation, 195
Simple Test of Goodness of Fit, 197
More on Tests of Goodness of Fit, 202

4.4. Stratification and Its Applications, 207
Basic Ideas, 208
Model and Implementation, 208
Analysis of Epidemiologic Matched Studies, 209
Homogeneity of Relative Risk in Matched Studies, 214

4.5. Nontime Application: Evaluation of Confounding Effects in ROC Studies, 215
Estimator for the ROC Function, 216
Use of Proportional Hazards Model, 217
Special Case, 219

Exercises, 221

Appendix A. Kidney Transplant Data 225

Appendix B. Hemodialysis Data 237

Appendix C. Ventilation Tube Data 245

References 249

Index 255

Preface

This book is intended to meet the need of practitioners and students in applied fields for a single, fairly thin volume covering major, updated methods in the analysis of survival data. It is written for the training of beginning graduate students in biostatistics, epidemiology, and environmental health, as well as for professional statisticians and biomedical research workers. As a book for professional statisticians, it is designed to offer sufficient details and a foundation to provide a better understanding of the various procedures as well as the relationships among different methods. However, the mathematics have been kept to a moderate level; most equations, derivations, and proofs involve only simple algebra and elementary calculus. As a book for students in applied fields and as a reference book for practicing biomedical research workers, *Applied Survival Analysis* is application oriented. It introduces applied research areas, a number of real-life examples and questions, most of which are completely solved. Those unfamiliar with calculus may choose to skip certain subsections without much discontinuity: Sections 3.5 and 4.5 may be skipped entirely.

The origin of survival analysis can be traced to early work on mortality tables, which was followed and expanded by statistical research for engineering applications. Most of the latter was concentrated on parametric models. Within the past few decades, there has been a great increase in the volume of medical research and clinical trials that has shifted the statistical focus to nonparametric methods. While this book attempts to cover both approaches, the major emphasis is on the more recent nonparametric developments with applicability to health research. Of course, the subjects covered in any text are to some extent a reflection of the author's own interests and contributions to the field, and this book is not an exception.

The book is divided into four chapters. Chapter 1 reviews some relevant basic methods in nonparametric statistics, introduces basic concepts—including the concept of censoring, which distinguishes survival analysis from other areas of statistics—and provides many examples. Chapter 2 covers some of the most widely used methods, both parametric and nonparametric, for estimating the survival function and other related parameters; some special health applications are also included. Chapter 3 is devoted to techniques for comparing survival distributions; topics include applications in the analysis of epidemiological cohort studies and sample size determination. Chapter 4 deals with the identification of prognostic and/or risk factors related to survival time. This last chapter covers standard procedures originating from Cox's proportional hazards model and includes recent topics such as goodness of fit and residual analysis. Appendix A provides a large real-life data set in end-stage renal diseases, which is used as a source for many examples and which can be used by instructors to create more exercises. This data set consists of 469 patients with kidney transplants; the primary endpoint was graft survival and time to graft failure was recorded in months. The study included measurements of many covariates that may be related to survival experience. Seven such variables are given in this data set: age (years) at transplant, sex, duration (days) of dialysis prior to transplant, diabetes, blood transfusion (blood units), mismatch score, and use of ALG (an immune suppression drug). Appendix B gives data for a group of 516 low-risk end-stage renal disease patients on hemodialysis. End-stage renal disease is characterized by an irreversible chronic renal failure. Major categories of treatment include renal transplantation (Appendix A), hemodialysis, and peritoneal dialysis. Hemodialysis achieves removal of toxins, electrolytes, and excess fluid via extracorporeal circulation of blood through a dialyzer (or external artificial kidney); treatments are usually scheduled for 3 to 4 hours, 3 times weekly. The low-risk patients in this data set consist of those without comorbidities such as arteriosclerotic heart disease, peripheral vascular disease, chronic obstructive pulmonary disease, cancer, or diabetes. The primary endpoint was patient survival and the above-mentioned comorbidities have been shown to have a significant effect on survival experience. In addition to survival time (or duration of dialysis, in months), this data set included age, sex, and race (white/nonwhite). Appendix C provides a set of paired survival data. Seventy-eight children, who received bilateral tympanostomy tubes as a treatment for their chronic

PREFACE

otitis media with effusion, were enrolled and randomized; half of the group received additional drug treatment after surgery. The aim of the medical treatment is to prolong the life of the tubes.

I would like to express my thanks to colleagues for their extensive comments and suggestions; the comments from many of my biostatistics graduate students at the University of Minnesota also have been most helpful. In the fall of 1980, a quarter in the second-year three-quarter course sequence for the masters program in biostatistics was devoted to the study of techniques used in analyzing survival data. It has been suggested by students and colleagues that it would be worthwhile writing and keeping notes for my lectures, and the contents herein represent that many-year effort. It was my own decision to keep this course applicational; our program has been supplemented by a second course devoted to the theories. However, the writing of this text was mostly completed before I started my work at The University of Minnesota Cancer Center; therefore, only some cancer data sets were included in this edition. With the new connection, I hope that the book will be greatly enriched in future editions.

CHAP T. LE

Edina, Minnesota
August 1997

CHAPTER 1

Basic Concepts in Survival Analysis

1.1. Study Designs
1.2. Survival Time
1.3. Functions of Survival Time
1.4. Regression Analysis
1.5. Examples of Survival Data
1.6. Examples of Survival Analysis
   Empirical Cumulative Distribution Function
   Comparison of Two Treatments
   Comparison of Several Treatments
   Rank Correlations
   Maximum-Likelihood Estimation (MLE)
1.7. A Special Problem: Incomplete Data
1.8. Data Presentation: Event Chart
1.9. Number of Deaths
1.10. Computational Implementation
Exercises

One of the major activities in health research is the identification of risk factors for diseases, for example, a study on the relationship between smoking and lung cancer. Such a relationship is proved (or disproved) by conducting a scientific investigation. Scientific investigations, like everything else in our society, involve uncertainty. Uncertainties arise because of variability in information (nature is complex, all methods—observational as well as experimental—are imperfect, there are always human biases and errors, moving targets) and its incompleteness (we deal with samples—not populations—due to cost and time constraints as well as our interests in future applications). Like everyone else in our society, when a scientist seeks to understand or explain something—the effect of an exposure or a risk factor for a disease—he or she usually

1. Proposes a hypothesis.
2. Seeks to test or investigate the proposed hypothesis by experiment or observation.
3. Decides, from information collected, if the hypothesis is strongly supported or estimates the extent of the effect (or effects) involved.

1.1. STUDY DESIGNS

In observing these various phenomena, the scientist is usually interested in observations on different characteristics. These characteristics are referred to as *variables*; the values of the observations recorded for them are referred to as *data*. Data from epidemiologic studies may come from different sources, with the two fundamental designs being *retrospective* and *prospective*.

Retrospective studies gather past data from selected cases and controls to determine differences, if any, in the exposure to a suspected risk factor. They are commonly referred to as *case-control studies*. Cases of a specific disease are ascertained as they arise from population-based disease registers or lists of hospital admissions, and controls are sampled either as disease-free individuals from the population at risk or as hospitalized patients having a diagnosis other than the one under study. The advantages of a retrospective study are that it is economical and it is possible to obtain answers to research questions relatively quickly because the cases are already available. Major limitations relate to the accuracy of exposure histories and uncertainty about the appropriateness of the control sample; these problems frequently hinder the precision of such studies and make them less preferred than prospective studies.

Prospective studies are usually cohort studies in which one enrolls a group of persons of comparable health and follows them over a certain period to observe the time at which a disease or another endpoint develops. Given the time-to-event data, the researcher then seeks to determine whether there is a statistical relation between an

SURVIVAL TIME

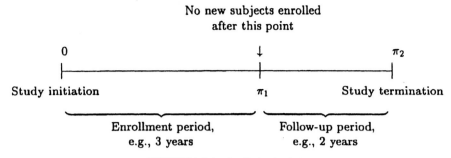

FIGURE 1.1. A clinical trial.

exposure and the disease. Another form of the prospective study consists of clinical trials. A clinical trial is a controlled experiment that usually seeks to determine the effectiveness of a new drug or treatment method. A controlled clinical trial (Figure 1.1) usually consists of two periods:

1. During the enrollment period, the interval $(0, \Pi_1)$, subjects enter the study sequentially and are randomized to receive either a potentially therapeutic agent or a placebo (or the standard therapeutic treatment). More than two treatment groups are possible.
2. Recruited subjects are then followed to the time of the primary ending event, or to time π_2 ($\pi_2 \geq \pi_1$); the interval (π_1, π_2) is called the *follow-up period.* In some study designs we may have $\pi_1 = \pi_2$. As an example, a study may consist of 3 years of enrollment and 2 years of follow-up ($\pi_1 = 3$ years, $\pi_2 = 5$ years); no patients are enrolled during the last 2 years.

1.2. SURVIVAL TIME

In prospective studies, the important feature is not only the outcome event, such as death, but the time to that event, the *survival time* (Figure 1.2). In order to determine the survival time T, three basic elements are needed:

1. Time origin or starting point
2. Ending event of interest
3. Measurement scale for the passage of time

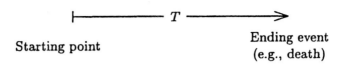

FIGURE 1.2. Survival time.

For example, the life span T from birth (starting point) to death (ending event) in years (measurement scale).

The time origin or starting point should be precisely defined, but it need not be birth; it could be the start of a new treatment (randomization date in a clinical trial) or the admission to a hospital or a nursing home. The ending event should also be precisely defined, but it need not be death; a nonfatal event such as the relapse of a disease (e.g., leukemia) or the relapse from a smoking cessation or the discharge to the community from a hospital or a nursing home satisfy the definition and are acceptable choices. The use of calendar time in health studies is common and meaningful; however, other choices for a time scale are justified—for example, hospital cost (in dollars) from admission (starting point) to discharge (ending event).

1.3. FUNCTIONS OF SURVIVAL TIME

The distribution of the survival time T from enrollment or starting point to the event of interest, considered as a random variable, is characterized by either one of two equivalent functions: the *survival function* and the *hazard function*.

The survival function, denoted $S(t)$, is defined as the probability that an individual survives longer than t units of time:

$$S(t) = \Pr(T > t)$$

$S(t)$ is also known as the *survival rate*; for example, if times are in years, then $S(2)$ is the 2-year survival rate, $S(5)$ is the 5-year survival rate, and so forth. The graph of $S(t)$ versus t is called the *survival curve*; a steep curve represents shorter survival and a gradual or flat curve represents longer survival. The survival function is also used to determine relevant parameters, such as the median

FUNCTIONS OF SURVIVAL TIME

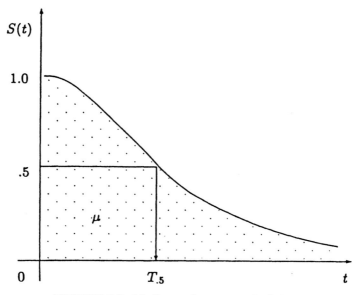

FIGURE 1.3. Median and mean survival time.

(and other percentiles) and mean survival times. The median $T_{.5}$ is a time point at which $S(T_{.5}) = .5$ (see Figure 1.3); in other words, exactly half of the population survives $T_{.5}$ units of time or longer. The mean μ is represented by the area under the survival curve; see Figure 1.3. Since most survival curves are positively skewed, $T_{.5} < \mu$.

The fact that the population mean μ is represented as the area under the survival curve can be easily seen as follows. In Figure 1.4, the shaded rectangle has (for very small time increment δ)

$$\text{Height} = S(t) - S(t + \delta)$$
$$= \Pr(T > t) - \Pr(T > t + \delta)$$
$$= \Pr[\text{die within } (t, t + \delta)]$$
$$\text{Width} = t$$

Therefore,

$$\text{Area} = t \Pr[\text{die within } (t, t + \delta)]$$

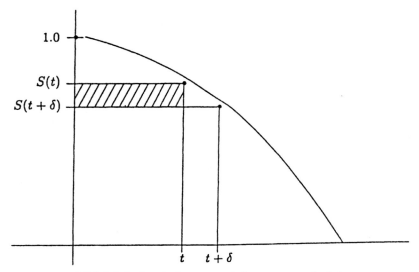

FIGURE 1.4. Survival curve and the mean survival time.

The area under the survival curve, which is the sum of the areas of these horizontal rectangles, is thus equal to

$$\text{Area under curve} = \sum t \Pr[\text{die within } (t, t+\delta)]$$

$$= \mu \quad (\text{as } \delta \to 0)$$

Usually the distribution of T, viewed as a random variable, is characterized by the *probability density function* $f(t)$ or the *cumulative distribution function* $F(t)$ defined by

$$f(t) = \lim_{\delta \downarrow 0} \frac{\Pr(t \leq T \leq t + \delta)}{\delta}$$

or

$$f(t)\,dt = \Pr(t \leq T \leq t + dt)$$

FUNCTIONS OF SURVIVAL TIME

and

$$F(t) = \Pr(T \leq t)$$
$$= \int_0^t f(x)\,dx$$
$$= 1 - S(t)$$

Using the probability density function, we can also show that μ equals *the area under the survival curve* by the following analytical proof:

$$\mu = E(T)$$
$$= \int_0^\infty t f(t)\,dt$$
$$= \int_0^\infty t\,d[1 - S(t)]$$
$$= \int_0^\infty \int_0^t dx\,d[1 - S(t)]$$
$$= \int_0^\infty \int_x^\infty d[1 - S(t)]\,dx$$
$$= \int_0^\infty S(x)\,dx$$

Similarly, given t, the mean residual life

$$r(t) = E(T - t \mid t \leq T)$$

can be expressed as

$$r(t) = \frac{1}{S(t)} \int_t^\infty S(x)\,dx$$

It can also be shown that the mean residual life function determines the distribution of T.

The *hazard* or *risk function* $\lambda(t)$ gives the *instantaneous* failure rate assuming that the individual has survived to time t,

$$\lambda(t) = \lim_{\delta \downarrow 0} \frac{\Pr(t \leq T \leq t + \delta \mid t \leq T)}{\delta}$$

or

$$\lambda(t)\,dt = \Pr(t \leq T \leq t + dt \mid t \leq T)$$

Thus, for a small time increment δ, the probability of an event occurring during time interval $(t, t + \delta)$ is given approximately by

$$\lambda(t) \cong \frac{\Pr(t \leq T \leq t + \delta \mid t \leq T)}{\delta}$$

In other words, the hazard or risk function $\lambda(t)$ approximates the proportion of subjects dying or having events per unit time around time t. Note that this differs from the density function represented by the usual histogram; in the case of $\lambda(t)$ the numerator is a *conditional* probability. $\lambda(t)$ is also known as the *force of mortality* and is a measure of the proneness to failure as a function of age of the individual. The hazard function may increase or decrease with time for long-term and short-term risks, respectively, or remain constant, or even indicate a more complicated process. For example, patients with acute leukemia who do not respond to treatment have an increasing hazard whereas the hazard function associated with organ transplant survival is decreasing. Complicated cases include the so-called bathtub curve describing the process of human life, and survival patterns with an initially increasing and then decreasing hazard rate; such examples include chronic leukemia and Hodgkin's disease (see Figure 1.5). When the hazard remains constant, we have the constant-risk (or exponential) model, which allows meaningful use of a statistic, the *rate*, frequently quoted in clinical research: deaths per 1000 treatment months (or person-years).

FUNCTIONS OF SURVIVAL TIME

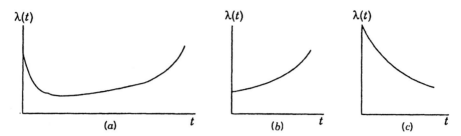

FIGURE 1.5. Some types of hazard functions: (*a*) hazard for human mortality, (*b*) positive aging, and (*c*) negative aging.

It can be seen that the hazard function is related to the others, viz.

$$\lambda(t) = f(t)/S(t)$$
$$= \left[-\frac{d}{dt}S(t)\right]\Big/S(t)$$

and by integrating and exponentiating both sides of this equation we have

$$\boxed{S(t) = \exp\left[-\int_0^t \lambda(x)\,dx\right]}$$

in which

$$H(t) = \int_0^t \lambda(x)\,dx$$

is called the *cumulative hazard* (up to time *t*). The following are a few typical *continuous* models.

Exponential Model

$$\boxed{\lambda(t) = \rho \qquad \text{a constant} \quad (\rho > 0)}$$

$$S(t) = e^{-\rho t} \qquad \text{for} \quad t > 0$$

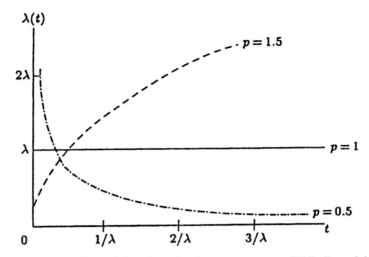

FIGURE 1.6. Hazard functions for the two-parameter Weibull model.

Weibull Model

$$\lambda(t) = \lambda p(\lambda t)^{p-1} \quad \text{for} \quad t > 0$$
$$\text{where} \quad \lambda, p > 0$$

$$S(t) = \exp[-(\lambda t)^p] \quad \text{for} \quad t > 0$$

The Weibull is a simple generalization of the exponential ($p = 1$) where the monotonicity of the hazard function is determined by the *shape parameter* p; $\lambda(t)$ is increasing if $p > 1$, decreasing if $p < 1$ (see Figure 1.6). Other models are Gompertz, log-logistic, and so forth (see Chapter 2).

If we consider T as a discrete random variable taking values $x_1 < x_2 < \cdots$ with associated probability mass

$$f(x_i) = \Pr(T = x_i)$$
$$= 0 \quad \text{elsewhere}$$

FUNCTIONS OF SURVIVAL TIME

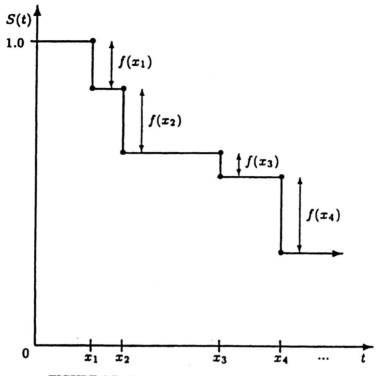

FIGURE 1.7. $S(t)$ is a step function for a discrete T.

then the survival function is a step function

$$S(t) = \sum_{j; t \leq x_j} f(x_j)$$

with jumps downward at $x_1, x_2, \ldots$ (see Figure 1.7). The hazard λ_i at x_i is still defined as the conditional probability of having an event at x_i, that is,

$$\lambda_i = \Pr[T = x_i \mid T \geq x_i]$$
$$= \frac{f(x_i)}{S(x_i)}$$

This leads to simple expressions for $f(t)$ and $S(t)$,

$$f(x_i) = \lambda_i \prod_{j=1}^{i-1}(1-\lambda_j)$$

$$S(x_i) = \prod_{j=1}^{i-1}(1-\lambda_j)$$

and

$$S(t) = \prod_{j:x_j<t}(1-\lambda_j) \quad \text{for} \quad t > 0$$

By nature, T is a continuous random variable; however, discrete models provide an efficient way to handle tied observations in the data.

1.4. REGRESSION ANALYSIS

When a population is subdivided into two subpopulations, E (exposed) and E' (nonexposed), by the presence or absence of a certain characteristic (an exposure such as smoking), each subpopulation corresponds to its own hazard or risk function and the ratio of two such functions:

$$\text{RR}(t) = \frac{\lambda(t;E)}{\lambda(t;E')}$$

is called the *relative risk* of exposure to factor E. In general, the relative risk $\text{RR}(t)$ is a function of time and measures the magnitude of an effect; when it remains constant, $\text{RR}(t) = \rho$, we have a *proportional hazards model* (PHM):

$$\lambda(t;E) = \rho\lambda(t;E')$$

with the risk of the nonexposed subpopulation served as the baseline. This is a multiplicative model; another way to express this model is

$$\lambda(t) = \lambda_0(t)e^{\beta x}$$

where $\lambda_0(t)$ is $\lambda(t; E')$—the hazard function of the unexposed subpopulation—and the indicator (or *covariate*) x is defined as

$$x = \begin{cases} 0 & \text{if unexposed} \\ 1 & \text{if exposed} \end{cases}$$

The "regression coefficient" β represents the relative risk on the log scale. Of course, this model works with any covariate X—continuous or categorical; the above binary covariate is only a very special case.

1.5. EXAMPLES OF SURVIVAL DATA

Example 1. A group of 12 hemophiliacs, all under 41 years of age at the time of HIV seroconversion, were followed from primary AIDS diagnosis until death (ideally we should take, as a starting point, the time at which an individual contracts AIDS rather than the time at which the patient is diagnosed, but this information is unavailable). Survival times (in months) from diagnosis until death of these hemophiliacs were: 2, 3, 6, 6, 7, 10, 15, 15, 16, 27, 30, and 32.

Example 2. Suppose that we are interested in studying patients with systemic cancer who subsequently develop a brain metastasis; our ultimate goal is to prolong their lives by controlling the disease. A sample of 23 such patients, all of whom were treated with radiotherapy, were followed from the first day of their treatment until recurrence of the original tumor. Recurrence is defined as the reappearance of a metastasis in exactly the same site, or, in the case of patients whose tumor never completely disappeared, enlargement of the original lesion. Times to recurrence (in weeks) for the 23 patients were: 2, 2, 2, 3, 4, 5, 5, 6, 7, 8, 9, 10, 14, 14, 18, 19, 20, 22, 22, 31, 33, 39, 195.

Example 3. The remission times (in weeks) of 42 patients with acute leukemia were reported in a clinical trial undertaken to assess the ability of 6–mercaptopurine (6–MP, a new drug) to maintain remission. Each patient was randomized to receive either 6–MP or a placebo; there were 21 patients in each group. The study was terminated after one year; data from placebo patients were complete: 1, 1, 2, 2, 3, 4, 4, 5, 5, 8, 8, 8, 8, 11, 11, 12, 12, 15, 17, 22, 23 (Freireich et al., 1983; Gehan, 1965a).

Example 4. A laboratory investigator interested in the relationship between diet and the development of tumors divided 90 rats into three groups and fed them with low-fat, saturated-, and unsaturated-fat diets, respectively. The rats were the same age and species and were in similar physical condition. An identical amount of tumor cells were injected into a foot pad of each rat. The rats were observed for 200 days; the following were tumor-free times (in days) of the rats fed with unsaturated-fat diet: 112, 68, 84, 109, 153, 143, 60, 70, 98, 164, 63, 63, 77, 91, 91, 66, 70, 77, 63, 66, 66, 94, 101, 105, 108, 112, 115, 126, 161, and 178. The tumor-free time is the time from injection of tumor cells to the time that a tumor develops; all 30 rats in the unsaturated-fat diet group developed tumors within 200 days (Lee, 1980).

Example 5. Data are shown below for two groups of patients who died of acute myelogenous leukemia. Patients were classified into the two groups according to the presence or absence of a morphologic characteristic of white cells. Patients termed AG-positive were identified by the presence of Auer rods and/or significant granulature of the leukemic cells in the bone marrow at diagnosis. For the AG-negative patients these factors were absent. Leukemia is a cancer characterized by an overproliferation of white blood cells; the higher the white blood count (WBC), the more severe the disease. Thus, when predicting a leukemia patient's survival time it is realistic to make a prediction dependent on WBC and any other covariates that are indicators of the progression of the disease; here we want to use as covariate $x = \ln[(WBC)]$.

(AG Positive) $N = 17$		(AG Negative) $N = 16$	
White Blood Count (WBC)	Survival Time (weeks)	White Blood Count (WBC)	Survival Time (weeks)
2,300	65	4,400	56
750	156	3,000	65
4,300	100	4,000	17
2,600	134	1,500	7
6,000	16	9,000	16
10,500	108	5,300	22
10,000	121	10,000	3
17,000	4	19,000	4
5,400	39	27,000	2
7,000	143	28,000	3
9,400	56	31,000	8
32,000	26	26,000	4
35,000	22	21,000	3
100,000	1	79,000	30
100,000	1	100,000	4
52,000	5	100,000	43
100,000	65		

Example 6. In a clinical trial for lung cancer treatment by the Veterans' Administration, males with advanced inoperable lung cancer were randomized to either a standard (0) or test (1) chemotherapy. The primary endpoint was time to death (in days). As is common in such studies, even with randomization, there was much heterogeneity between patients in, for example, disease extent and pathology, previous treatment of the disease, demographic background, and initial health status. The data below include a measure at randomization, x_1, of the patients' performance status or "Karnofsky rating": 10–30 completely hospitalized, 40–60 partial confinement, 70–90 able to care for self. Also available were time in months from diagnosis to randomization (x_2), age in years (x_3),

prior therapy (x_4; 0 = no, 1 = yes), and histological type of tumor: squamous, small cell, adeno, and large cell. Only data for one subgroup (test treatment, large cell) are given here (Kalbfleisch and Prentice, 1980).

t	x_1	x_2	x_3	x_4
52	60	4	45	0
164	70	15	68	1
19	30	4	39	1
53	60	12	66	0
15	30	5	63	0
43	60	11	49	1
340	80	10	64	1
133	75	1	65	0
111	60	5	64	0
231	70	18	67	1
378	80	4	65	0
49	30	3	37	0

1.6. EXAMPLES OF SURVIVAL ANALYSIS

Many statistical methods depend on assuming specific parametric model(s) for the distribution(s) of the data. For example, derivations of popular procedures such as the two-sample t test and the one-way Analysis of Variance (ANOVA) are based on the assumption that data are normally distributed. Of course, these procedures are supported by the central limit theorem; but others, such as the comparison of two exponential distributions, are not that robust. That is, they are sensitive to distributional assumptions; departures would have effects on the results. Moreover, all methods—including those supported by the central limit theorem—are sensitive to extreme observations, that is, a few very small or very large (possibly erroneous) data values. To handle problems in which the various assumptions underlying procedures cannot be met, statisticians have developed many alternate techniques that have become known as *nonparametric methods* or *statistical methods based on ranks*. These methods have the additional advantage that they are less sensitive with respect to extreme observations.

Empirical Cumulative Distribution Function

Let X be a random variable. The cumulative distribution function (cdf) is defined by

$$F(x) = \Pr(X \leq x)$$

Given a random sample

$$\{x_i\}_{i=1}^n$$

the empirical cdf is defined as

$$\hat{F}_n(x) = (\text{number of } x_i\text{'s} \leq x)/n$$

Since $n\hat{F}_n(x)$ has a binomial distribution $B(n,p)$ with

$$p = F(x)$$

$\hat{F}_n(x)$ is an unbiased estimate of $F(x)$ with variance

$$F(x)[1 - F(x)]/n$$

which is estimated by

$$\hat{F}_n(x)[1 - \hat{F}_n(x)]/n$$

Moreover, the estimate is consistent, that is, $\hat{F}_n(x)$ converges in probability to $F(x)$, and $\hat{F}_n(x)$ is asymptotically normal for each x. This estimator has a number of important applications:

(i) The Kolmogorov statistic

$$D_n = \sup_x |\hat{F}_n(x) - F_0(x)|$$

has been used for goodness-of-fit testing against

$$\mathcal{H}_0 : F = F_0$$

(ii) The Kolmogorov–Smirnov statistic

$$D_{mn} = \sup_x |\hat{F}_n(x) - \hat{G}_m(x)|$$

has been used for testing

$$\mathcal{H}_0 : F(x) = G(x)$$

using two independent samples $\{x_i\}_{i=1}^n$ and $\{y_j\}_{j=1}^m$.

(iii) An estimate of the population median, $\hat{x}_{.5}$, can be obtained from

$$\hat{F}_n(\hat{x}_{.5}) = .5$$

Other quantiles (e.g, 25th and 75th percentiles) can be obtained similarly.

We can use these results for the empirical cdf $\hat{F}_n$ by simply noting that the survival function $S(t)$ can be expressed as

$$S(t) = 1 - F(t)$$

In other words, if we define for a sample $\{t_i\}$, $1 \leq i \leq n$:

$$\hat{S}_n(t) = (\text{number of } t_i\text{'s} > t)/n$$

then $\hat{S}_n(t)$ is an unbiased estimate of $S(t)$ with variance

$$S(t)[1 - S(t)]/n$$

which is estimated by

$$\hat{S}_n(t)[1 - \hat{S}_n(t)]/n$$

Example 7. Consider the data on the 12 hemophiliacs of Example 1. What are we able to infer about the survival of the pop-

EXAMPLES OF SURVIVAL ANALYSIS

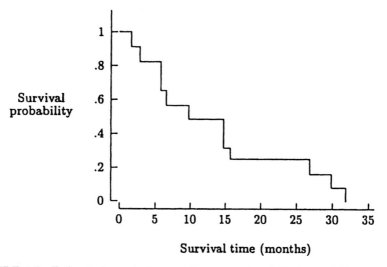

FIGURE 1.8. Estimated survival curve for a sample of 12 hemophiliac patients less than 41 years of age at HIV seroconversion.

ulation of hemophiliacs on the basis of these 12 individuals in the sample? Using the above method, we can obtain:

Time t	$\hat{S}_n(t)$
0	1.0000
2	0.9167
3	0.8333
6	0.6667
7	0.5833
10	0.5000
15	0.3333
16	0.2500
27	0.1667
30	0.0833
32	0.0000

The corresponding curve is plotted in Figure 1.8. Since $\hat{S}_n(t)$ may be denoted by a simpler notation $\hat{S}(t)$, calculated using the data in a sample under observation, it is merely an estimate of the true population survival function $S(t)$. To quantify the sampling variability involved, we can calculate the standard error of $\hat{S}(t)$

$$\left\{\frac{\hat{S}(t)[1-\hat{S}(t)]}{n}\right\}^{1/2}$$

and construct a pair of confidence bands around the estimated survival curve of Figure 1.8 (more details on confidence intervals in Chapter 2).

Comparison of Two Treatments

The problem of comparing two treatments arises in many different contexts. Does the new "cure" prolong the life of cancer patients? Is the harmful effect of cigarettes reduced by filtering? Typically, for comparing a new treatment or procedure with a standard, $(n_1 + n_2)$ subjects are divided at random into a group of n_1 who will receive the new treatment and a control group of n_2 who will be treated by the standard method.

Wilcoxon Rank-Sum Test

This is perhaps the most popular nonparametric procedure. At the termination of the study, the subjects in both groups are *jointly ranked* according to some response that measures the success of the treatment, such as survival time. If the difference between the two treatments is sufficiently strong, it will typically be reflected in the ranks. Thus, to complete the process, it is only necessary to decide just when the treatment ranks $\{R_1, R_2, \ldots R_{n_1}\}$ are sufficiently large. A simple and effective statistic is the Wilcoxon's rank sum (Wilcoxon, 1945)

$$\boxed{W = \sum_{i=1}^{n_1} R_i}$$

and the null hypothesis $\mathcal{H}_0$ is rejected (and the superiority of the new treatment acknowledged) when W is sufficiently large. Under $\mathcal{H}_0$, we have

$$E_0(W) = \tfrac{1}{2}n_1(n_1 + n_2 + 1)$$

and
$$\text{Var}_0(W) = \tfrac{1}{12}n_1 n_2(n_1 + n_2 + 1).$$

and the significance level can be obtained by referring the standardized statistic

$$z = \frac{W - E_0(W)}{\{\text{Var}_0(W)\}^{1/2}}$$

to the standard normal distribution when the sample sizes n_1, n_2 are relatively large (say, $n_1 \geq 10$, $n_2 \geq 10$).

Mann–Whitney Version

Denote the treatment and control observations by

$$\{x_i\}_{i=1}^{n_1} \quad \text{and} \quad \{y_j\}_{j=1}^{n_2}$$

respectively, and consider all possible pairs of observations (x_i, y_j) consisting of one x and one y. Among $n_1 n_2$ such pairs there will be some for which $y_j < x_i$ and some with $x_i < y_j$. Let

$$\boxed{W_{YX} = \text{number of pairs } (x_i, y_j) \text{ with } y_j < x_i}$$

as it turns out, we have

$$W_{YX} = W - \tfrac{1}{2}n_1(n_1 + 1)$$

where W is the above Wilcoxon rank-sum statistic. This means that the use of W_{YX} is equivalent to the use of the Wilcoxon rank-sum statistic W; W_{YX} is known as the *Mann–Whitney statistic*. We have

$$E_0(W_{YX}) = \tfrac{1}{2}n_1 n_2$$

and

$$\begin{aligned}\text{Var}_0(W_{YX}) &= \text{Var}_0(W) \\ &= \tfrac{1}{12}n_1 n_2(n_1 + n_2 + 1)\end{aligned}$$

Siegel–Tukey Test

In the formation of the Wilcoxon/Mann–Whitney test, it was implicitly assumed that the effect of the new treatment, if any, is to increase (or decrease) the responses as compared to what they would have been without the treatment. However, this is not the only alternative to the null hypothesis of no treatment effects. Some treatments may be associated with responses in a nonmonotonic pattern; some do not change the average response but resulting responses will be less dispersed about this average value. To obtain a suitable nonparametric test, note that treatment responses are more likely to be in the center part whereas control responses tend to be further away when both groups are jointly ranked. The Siegel–Tukey test uses a ranking scheme as follows: assign rank 1 to the smallest observation, rank 2 to the largest, rank 3 to the second largest, rank 4 to the second smallest, rank 5 to the third smallest observation, and so on as indicated in this figure ($n_1 + n_2 = 9$):

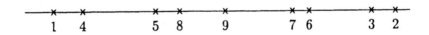

In view of the random assignment of ranks under $\mathcal{H}_0$, the rank-sum statistic would have the same distribution as in the Wilcoxon test, even with this new definition of ranks; large values of W are in favor of the new treatment (smaller dispersion).

An alternative to the Siegel–Tukey test is an application of the Wilcoxon test to compare the deviations from the means:

$$\{d_i = |x_i - \bar{x}|\}_{i=1}^{n_1} \quad \text{vs.} \quad \{d_j = |y_j - \bar{y}|\}_{j=1}^{n_2}$$

(Levene, 1960).

Example 8. Consider the data for lung cancer patients in Example 6 and divide the sample of survival times into two groups:

Group 1: no prior therapy   52, 53, 15, 133, 111, 378, 49
Group 2: with prior therapy   164, 19, 43, 340, 231

In order to apply the Wilcoxon test, we rank the data jointly as follows:

Group: (1) 15   49 52 53 111 133   378
          (2)   19 43   231 240 340
Rank: 1 2 3 4 5 6 7 8   9 10 11 12

The sum of the ranks for group 2 is: $W = 2+3+9+10+11 = 35$ leading to a z score of

$$z = \frac{35 - (5)(13)/2}{\{(5)(7)(13)/12\}^{1/2}}$$

$$= .41$$

(the sample sizes may be too small for a normal approximation).

Comparison of Several Treatments

Comparative studies frequently involve the simultaneous comparison not just of two but of three or more treatments or conditions; one may, for example, wish to compare the effect on growth of several diets. Although somewhat more complex, such comparisons share many of the features found in the comparison of only two treatments.

Kruskal–Wallis Test

This test can be considered a direct extension of the Wilcoxon rank-sum test. Suppose that N subjects are available for a comparison of s treatments and that it is decided to allocate n_i subjects to the ith treatment so that

$$N = \sum n_i$$

with the subjects being assigned to the treatments at random. At the termination of the study, the N subjects are jointly ranked according to the values of some response.

Let R_{ij} be the rank of the jth subject in group i and

$$\overline{R}_i = \left(\sum_j R_{ij}\right)\bigg/ n_i$$

be the average rank for the ith treatment. If the treatments differ widely from each other, one would expect big differences among the values of the $\overline{R}_i$'s. On the other hand, when the null hypothesis $\mathcal{H}_0$ of no difference is true, the $\overline{R}_i$'s may be expected to be close to each other and hence also close to the overall average

$$\overline{R} = \left(\sum_{i,j} R_{ij}\right)\bigg/ N$$
$$= (N+1)/2$$

A convenient criterion for measuring the overall closeness of the $\overline{R}_i$'s to $\overline{R}$ is a weighted sum of squared differences leading to the Kruskal–Wallis statistic:

$$\boxed{K = \frac{12}{N(N+1)} \sum_{i=1}^{s} n_i \left(\overline{R}_i - \frac{N+1}{2}\right)^2}$$

$$= \frac{12}{N(N+1)} \sum n_i \overline{R}_i^2 - 3(N+1)$$

Under $\mathcal{H}_0$, the Kruskal–Wallis statistic K is distributed as chi-square with df $= s - 1$ degrees of freedom, and the null hypothesis $\mathcal{H}_0$ is rejected for large values of K.

Tests Against an Ordered Alternative

In the comparison of s treatments or conditions using the Kruskal–Wallis test, we reject the null hypothesis if and only if there is strong enough evidence to support the alternative hypothesis, that is, any difference among the s treatments. Suppose now that the treatments are ordered (of course, before the responses have been observed) in such a way that under the alternative to the null hypothesis $\mathcal{H}_0$, one would expect larger responses under treatment 2 than under treatment 1, under treatment 3 than under treatment 2, and so on (i.e., ordered alternatives). A test such as the Kruskal–Wallis test is no longer appropriate since it rejects $\mathcal{H}_0$ whenever the rank averages $\overline{R}_i$'s are sufficiently different, regardless of their order, while in the present case only an increasing trend in the sequence of $\overline{R}_i$'s supports the alternative over the null hypothesis.

Jonckheere Test

Let W_{ij} be the Mann–Whitney statistic comparing the ith and the jth treatments, that is, the number of pairs (α, β) for which $x_{i\alpha} < y_{j\beta}$. A test statistic against ordered alternatives proposed by Jonckheere (1954) is then the sum of the W_{ij}'s:

$$J = \sum_{i<j} W_{ij}$$

the null hypothesis being rejected for large values of J. For large values of N, the Jonckheere statistic J is approximately normally distributed with

$$E_0(J) = \frac{N^2 - \sum n_i^2}{4}$$

and

$$\mathrm{Var}_0(J) = \frac{N^2(2N+3) - \sum n_i^2(2n_i+3)}{72}$$

Cuzick Test

The Cuzick test statistic (1985) is as follows:

$$T = \sum_{i=1}^{N} z_i r_i$$

where N is the total number of observations in the combined sample, r_i is the rank of the ith observation in the combined sample, and z_i is a "score" associated with the group to which the ith observation belongs. In order to apply this test, the k samples of observations must be combined together and each observation ranked in magnitude relative to the other observations in the combined group. The rank for each observation is then assigned back to its original sample and the test statistic, T, is calculated. Asymptotically, under $\mathcal{H}_0$, the distribution of T is normal. The mean of T is

$$E(T) = \left(\sum_{j=1}^{N} j\right) E(Z) = (\tfrac{1}{2})N(N+1)E(Z)$$

where

$$E(Z) = \sum_{j=1}^{k} z_j p_j$$

k denotes the number of groups, z_j is equal to j, and p_j is the proportion of all observations that came from group j, that is,

$$p_j = n_j/N$$

The variance of T is

$$\mathrm{Var}(T) = [N^2(N+1)/12]\mathrm{Var}(Z)$$

where

$$\mathrm{Var}(Z) = \sum_{i=1}^{k} z_i^2 p_i - [E(Z)]^2$$

Le Test

The Le test (1988a) is derived from Spearman's ρ and is obtainable by a decomposition of the Kruskal–Wallis test. Le's test statistic is calculated using the combined sample ranks of the data, similar to the Cuzick test, and is given as

$$\boxed{W = \sum_{i=1}^{k} n_i(L_i - M_i)\overline{R}_i}$$

Here, n_i is the size of sample i; L_i is equal to the total number of observations in all the groups to the left of the ith group, with $L_1 = 0$; M_i is equal to the total number of observations in all the groups to the right of the ith group, with $M_k = 0$; and $\bar{R}_i$ is the average rank value for group i.

Under $\mathcal{H}_0$, W is approximately normally distributed with mean 0 and variance

$$\text{Var}(W) = [N(N+1)/12]\sum_{i=1}^{k} n_i(L_i - M_i)^2$$

where n_i, L_i, and M_i are the same as previously defined. The difference

$$K - W^2/\text{Var}(W)$$

[chi-square at $(k-2)$ degrees of freedom] can be used for testing departure from the trend.

Example 9. Consider the survival times of AG-positive patients in Example 5 and divide this sample into 3 groups:

Group 1: Those patients with WBC $\geq$ 100,000: 1, 1, 65
Group 2: Those patients with 10,000 $\leq$ WBC $<$ 100,000: 108, 121, 4, 26, 22, 5
Group 3: Those patients with WBC $<$ 10,000: 65, 156, 100, 134, 16, 39, 143, 56

In attempting to rank the data with ties, we assign the average of the ranks to each tie; results are as follows:

Group:
(1) 1 1   65
(2)   4 5   22 26   108 121
(3)   16   39 56 65 100   134 143 156
Rank:
     1.5 1.5 3 4   5   6   7   8   9 10.5 12   13 14   15   16   17

$$\bar{R}_1 = 4.5, \quad \bar{R}_2 = 7.83, \quad \text{and} \quad \bar{R}_3 = 11.56$$

An application of the Kruskal–Wallis test yields

$$K = \frac{12}{(17)(18)}[(3)(4.5)^2 + (6)(7.83)^2 + (8)(11.56)^2] - (3)(18)$$

$$= 4.73 \quad \text{at 2 dfs (degrees of freedom)} \quad (p = .0939)$$

An application of the Le test yields

$$W = (3)(0 - 14)(4.5) + (6)(3 - 8)(7.83) + (8)(9 - 0)(11.56)$$

$$= 408.42$$

leading to a z score of

$$z = 408.4 \bigg/ \left\{ \frac{(17)(18)}{12}[(3)(0-14)^2 + (6)(3-8)^2 + (8)(9-0)^2] \right\}^{1/2}$$

$$= 2.17 \quad \text{(one-tailed } p = .0150\text{)}$$

Rank Correlations

To measure the strength of a relationship between two random variables X and Y, we often use the coefficient of correlation R defined by

$$\boxed{R = \frac{\text{Cov}(X,Y)}{\sqrt{\text{Var}(X)\text{Var}(Y)}}}$$

with

$$\text{Var}(X) = E(X - \mu_X)^2$$
$$\text{Var}(Y) = E(Y - \mu_Y)^2$$

and

$$\text{Cov}(X,Y) = E[(X - \mu_X)(Y - \mu_Y)]$$

Suppose we observe these two random variables on each member of a sample of size n and let (x_i, y_i) be the observations on the ith individual. We then have, as an estimate of the coefficient of correlation R, the sample coefficient of correlation r,

$$r = \frac{\sum(x_i - \bar{x})(y_i - \bar{y})}{\{[\sum(x_i - \bar{x})^2][\sum(y_i - \bar{y})^2]\}^{1/2}} \qquad -1 \leq r \leq 1$$

Similar to the case of other parametric statistics/procedures, this measure of correlation is sensitive with respect to extreme observations.

Spearman's Rho

To form a nonparametric correlation measure, we rank the x_i's the y_i's separately and let $\{R_i\}$ and $\{S_i\}$ be their ranks, respectively. If we replace x_i by its rank R_i and y_i by its rank S_i in the formula for the sample correlation coefficient r, we arrive at the Spearman rank correlation coefficient rho (ρ):

$$\rho = \frac{\sum(R_i - \bar{R})(S_i - \bar{S})}{\{[\sum(R_i - \bar{R})^2][\sum(S_i - \bar{S})^2]\}^{1/2}}$$

Since the R_i's and the S_i's are each permutations of the first n integers $\{1, 2, \ldots, n\}$, we have

$$\bar{R} = \bar{S}$$
$$= (n+1)/2$$

Hence

$$\sum(R_i - \bar{R})^2 = \sum(S_i - \bar{S})^2$$
$$= \sum\left(i - \frac{n+1}{2}\right)^2$$
$$= \frac{n(n^2 - 1)}{12}$$

Also, we can write

$$\sum(R_i - \bar{R})(S_i - \bar{S}) = \sum R_i S_i - \frac{n(n+1)^2}{4}$$

$$= \frac{n(n^2-1)}{12} - \frac{1}{2}\sum(R_i - S_i)^2$$

Therefore the above Spearman rho can be simplified to

$$\rho = 1 - \frac{6\sum(R_i - S_i)^2}{n(n^2-1)} \qquad -1 \le \rho \le 1$$

(Of course, the sum $\sum(R_i - S_i)^2$ can be used as a statistic itself.)

Kendall's Tau

Kendall's tau is another nonparametric correlation measure that is based on the counts of concordant and discordant pairs of subjects. Two subjects i and j, with measurements (x_i, y_i) and (x_j, y_j), form a concordant pair if

$$(x_i - x_j)(y_i - y_j) > 0$$

or a discordant pair if

$$(x_i - x_j)(y_i - y_j) < 0$$

Let C and D be the number of concordant pairs and discordant pairs, respectively. Then the Kendall tau is defined by

$$\tau = \frac{C - D}{\frac{1}{2}n(n-1)} \qquad -1 \le \tau \le 1$$

As compared to the Spearman rho, Kendall's tau can be updated more easily if more data are collected and, for smaller values of n, its sampling distribution approximates the normal distribution more closely.

EXAMPLES OF SURVIVAL ANALYSIS

$2 \times k$ *Contingency Tables*

The observable responses in a comparative study need not be numerical but may specify only to which of a number of categories each subject belongs. Such data are described as categorical and in many practical solutions, the categorical response is ordinal; for example, the subject is observed to be young, middle-aged, or old, or the subject receives a grade of A, B, C, D, or F in a course. In these cases the data may be represented by a $2 \times k$ contingency table.

Status	\multicolumn{5}{c}{Response level}	Total

Status	1	...	i	...	k	Total
Death	n_{11}	...	n_{1i}	...	n_{1k}	n_1
Alive	n_{21}	...	n_{2i}	...	n_{2k}	n_2
Total	m_1	...	m_i	...	m_k	N

From the frequencies in this contingency table, the number of concordant pairs C and the number of discordant pairs D are given by

$$C = \sum_{i=1}^{k-1} n_{1i} \left[\sum_{j=i+1}^{k} n_{2j} \right]$$

and

$$D = \sum_{i=1}^{k-1} n_{2i} \left[\sum_{j=i+1}^{k} n_{1j} \right]$$

These statistics have a number of important applications.

(i) The difference

$$W = C - D$$

has been used for testing against the null hypothesis of no difference (between the two treatments). Under $\mathcal{H}_0$, W is asymptotically normally distributed with

$$E_0(W) = 0$$

$$\text{Var}_0(W) = \frac{n_1 n_2 \left[N^3 - \sum m_i^3\right]}{3N(N-1)}$$

(ii) Treatment effect (i.e., the relationship between the "column variable" and "row variable" in the above $2 \times k$ contingency table) can be measured using either

$$\gamma = \frac{C - D}{C + D}$$

or

$$\theta = \frac{C}{D}$$

Gamma (γ) acts as a coefficient of correlation ($-1 \leq \gamma \leq 1$) whereas theta (θ) can be treated as a generalized odds ratio.

Later in Chapter 3, we will apply these ideas for $2 \times k$-ordered contingency tables to the analysis of survival data.

Example 10. Consider again the data for AG-positive patients in Example 5 and suppose we want to investigate the relationship between survival time and white blood count (WBC). In order to calculate the Kendall τ, we proceed as follows:

1. First, the WBCs and times are presented in two columns, with the WBC values in the first column arranged from smallest to largest.
2. For each survival time in the second column, we count
   (i) The number of larger times and larger WBC (below it). The sum of these is C.
   (ii) The number of smaller times and larger WBC (below it). The sum of these is D.

From the following table, we have

$$\tau = \frac{33-101}{(17)(16)/2}$$

$$= -.50$$

indicating a moderate negative relationship.

WBC	Time	(i)	(ii)
750	156	0	16
2,300	65	5	9
2,600	134	1	13
4,300	100	3	10
5,400	39	5	7
6,000	16	7	4
7,000	143	0	10
9,400	56	3	6
10,000	121	0	8
10,500	108	0	7
17,000	4	4	2
32,000	26	1	4
35,000	22	1	3
52,000	5	1	2
100,000	1	1	0
100,000	1	1	0
100,000	65	0	0
	Totals	33	101

Maximum-Likelihood Estimation (MLE)

Suppose that we may be able to assume a parametric model for survival times, for example, an exponential model with density $f(t;\rho)$ or a Weibull model with density $f(t;\lambda,p)$.

Given a random sample of survival times

$$\{t_i\}_{i=1}^n$$

and a density function, denoted $f(t;\theta)$, the likelihood function of θ is

$$L(\theta) = \prod_{i=1}^{n} f(t_i;\theta)$$

and data can be analyzed using standard methods (MLE, score statistic, likelihood ratio statistic) associated with large-sample likelihood theory.

Example 11. For the exponential model

$$L = \prod_{i=1}^{n} \lambda e^{-\lambda t_i}$$
$$= \lambda^n e^{-\lambda \sum t_i}$$
$$= \lambda^n e^{-\lambda T} \quad \text{with} \quad T = \sum t_i \text{ being total time}$$

this leads to

$$\ln L = n \ln \lambda - \lambda T$$

so that from

$$0 = \frac{d}{d\lambda} \ln L$$
$$= \frac{n}{\lambda} - T$$

we have

$$\hat{\lambda} = \frac{n}{T}$$

(the usual *rate* frequently quoted in clinical research: deaths per treatment month or treatment year). From

$$-\frac{d^2}{d\lambda^2} \ln L = \frac{n}{\lambda^2}$$

the standard error (SE) of $\hat{\lambda}$ is given by

$$SE(\hat{\lambda}) = \frac{\hat{\lambda}}{\sqrt{n}}$$

For example, consider the times to recurrence (in weeks) for the 23 cancer patients of Example 2:

$$n = 23$$
$$T = 490$$

leading to

$$\hat{\lambda} = .047$$
$$SE(\hat{\lambda}) = .010$$

or a 95 percent confidence interval for λ: (.028, .066) or a 95 percent confidence interval (15.2 weeks, 35.7 weeks) for mean survival time ($\mu = 1/\lambda$).

1.7. SPECIAL PROBLEM: INCOMPLETE DATA

A special source of difficulty in the analysis of survival data is the possibility that some individuals may not be observed for the full time to failure or event. The so-called random censoring arises in medical applications with animal studies, epidemiological applications with human studies, or clinical trials. In these cases, observation is terminated before the occurrence of the event. In a clinical trial, for example, patients may enter the study at different times; then each is treated with one of several possible therapies after a randomization. We want to observe their lifetimes from enrollment, but censoring may occur in one of the following forms:

- Loss to follow-up—the patient may decide to move elsewhere
- Dropout—a therapy may have such bad effects that it is necessary to discontinue the treatment

- Termination of the study (for data analysis at a predetermined time)
- Death due to a cause not under investigation (e.g., suicide)

To make it possible for statistical analysis we make the crucial assumption that, conditionally on the values of any explanatory variables (or covariates), the prognosis for any individual who has survived to a certain time t should not be affected if the individual is censored at t. That is, an individual who is censored at t should be representative of all those subjects with the same values of the explanatory variables who survive to t. In other words, survival condition and reason of loss are independent; under this assumption, there is no need to distinguish the above four forms of censoring.

It can be seen in the context described above that the time origin need not be and usually is not the same calendar time for each individual. Most clinical trials have staggered entry, so that patients enter anytime over the enrollment period $(0, \Pi_1)$. Each patient's survival time is measured from his/her own date of entry. Figure 1.9 illustrates this concept; the above assumption on the independence between patient condition and reason for censoring allows us to put all observed times at the same makeup entry date ($t = 0$) as in part (b) of the figure.

We assume that observations available on the failure time of n individuals are usually taken to be independent. At the end of the study, our sample consists of n pairs of numbers (t_i, δ_i). Here δ_i is an indicator variable for survival status ($\delta_i = 0$ if the ith individual is censored; $\delta_i = 1$ if the ith individual failed) and t_i is the time to failure/event (if $\delta_i = 1$) or the censoring time (if $\delta_i = 0$); t_i is also called the *duration time*. We may also consider, in addition to t_i and δ_i, $(x_{1i}, x_{2i}, \ldots, x_{ki})$, a set of k covariates associated with the ith individual representing cofactors such as age, sex, treatment, and the like.

Example 12. In Example 3, we consider a study in which 42 leukemia patients were randomized to receive either the drug 6-MP or a placebo, 21 patients in each group. The remission times (in weeks) were as follows:

Placebo group: 1, 2, 2, 2, 3, 4, 4, 5, 5, 8, 8, 8, 8, 11, 11, 12, 12, 15, 17, 22, 23

SPECIAL PROBLEM: INCOMPLETE DATA

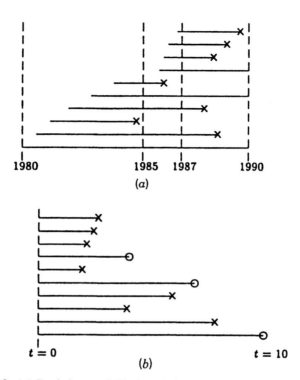

FIGURE 1.9. (a) Real time and (b) time T from entry (×, death; ○, censoring).

6–MP group: 6, 6, 6, 7, 10, 13, 16, 22, 23, 6^+, 9^+, 10^+, 11^+, 17^+, 19^+, 20^+, 25^+, 32^+, 32^+, 34^+, 35^+

in which a t^+ denotes a censored observation ($\delta = 0$), that is, a patient who was censored (by study termination) after t weeks without a relapse. For example, "10^+" is a case enrolled 10 weeks before study termination and still in remission at termination.

Example 13. In Example 4, we described a diet study and tumor-free times were given for the 30 rats fed with unsaturated-fat diet. Tumor-free times (days) for the other two groups are as follows:

Low-fat: 140, 177, 50, 65, 86, 153, 181, 191, 77, 84, 87, 56, 66, 73, 119, 140^+ and 14 rats at 200^+

Saturated-fat: 124, 58, 56, 68, 79, 89, 107, 86, 142, 110, 96, 142, 86, 75, 117, 98, 105, 126, 43, 46, 81, 133, 165, 170$^+$ and 6 rats at 200$^+$

(140$^+$ and 170$^+$ were due to accidental deaths without evidence of tumor.)

Example 14. In the clinical trial for lung cancer treatment by the Veteran's Administration (Example 6), there were several histological types of tumor. The following data were obtained from two other subgroups: (standard treatment, large cell) and (standard treatment, squamous cell).

Standard, large					Standard, squamous				
t	x_1	x_2	x_3	x_4	t	x_1	x_2	x_3	x_4
72	60	7	69	0	177	50	16	66	1
411	70	5	64	1	162	80	5	62	0
228	60	3	38	0	216	50	15	52	0
126	60	9	63	1	553	70	2	47	0
118	70	11	65	1	278	60	12	63	0
10	20	5	49	0	12	40	12	68	1
82	40	10	69	1	260	80	5	45	0
314$^+$	50	18	43	0	200	80	12	41	1
42	70	6	70	0	156	70	2	66	0
42	60	4	81	0	182$^+$	90	2	62	0
8	40	58	63	1	143	90	8	60	0
144	30	4	63	0	105	80	11	66	0
25$^+$	80	9	52	1	103	80	5	38	0
11	70	11	48	1	250	70	8	53	1
					100	60	13	37	1

There were two censored cases in the large-cell group (314$^+$ and 25$^+$) and one censored case in the squamous-cell group (182$^+$).

Example 15. Appendix A gives the data for 469 patients with kidney transplants. The primary endpoint was graft survival and time to graft failure was recorded in months. This study included measurements of many covariates that may be related to survival experience. Seven such variables are given in this data set: age (years) at transplant, sex, duration (days) of dialysis prior to transplant, diabetes, blood transfusion (blood units), mismatch score, and use of ALG (an immune suppression drug). Again, the need to accommodate censoring is important.

Example 16. Appendix B gives the data for a group of 516 low-risk end-stage renal disease patients on hemodialysis. End-stage renal disease is characterized by an irreversible chronic renal failure. Major categories of treatment include renal transplantation (see Example 15), hemodialysis, and peritoneal dialysis. Hemodialysis achieves removal of toxins, electrolytes, and excess fluid via extracorporeal circulation of blood through a dialyzer (or external artificial kidney); treatments are usually scheduled for 3 to 4 hours, 3 times weekly. The low-risk patients in this data set consist of those without comorbidities such as arteriosclerotic heart disease, peripheral vascular disease, chronic obstructive pulmonary disease, cancer, or diabetes. The primary endpoint was patient survival and the above-mentioned comorbidities have been shown to have a significant effect on survival experience. In addition to survival time (or duration of dialysis, in months), this data set included age, sex, and race (white/nonwhite).

Example 17. Appendix C gives a real set of paired survival data. The study enrolled 78 children with chronic otitis media with effusion (ear infection) who were having therapeutic myringotomy for bilateral tympanostomy tube placement and survival times (in weeks) are for the duration of the ventilating tubes. The data also included children from two different treatments.

1.8. DATA PRESENTATION: EVENT CHART

In the presence of censoring, basic graphical methods such as the histogram can no longer be used for data presentation. A new mode for presenting survival data has been proposed, called the *event chart*. (Goldman, 1992). A basic event chart is a dated diagram of death and censoring events of individual patients. Consider the simple example in Exercise 1.3; this data set is graphically presented in Figure 1.10.

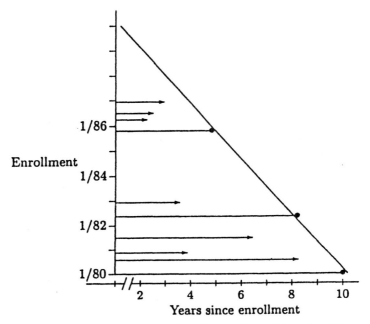

FIGURE 1.10. Basic event chart for the data of Exercise 1.3.

The vertical axis indicates the enrollment date, with a made-up entry date $t = 0$, and the horizontal axis represents the elapsed time since enrollment. Each horizontal or *event bar* is an individual record, graphed in ascending order of enrollment. An arrowhead at the end of a bar indicates time of death. The slanted dashed line represents the time of analysis (or study termination); at this point all living patients were censored, the censoring due to study termination. If any patients were lost to follow-up, the time of loss is indicated by an open circle at the end of the bar. This graphical tool has several advantages:

(i) The dates and timing of the events are retained. Therefore, it is possible to monitor a study for secular trends in accrual, loss, and event times. If there are excessive losses to follow-up, for example, the chart may help to diagnose the problem. Secular clustering of events may signal an epidemic.

(ii) The slanted termination line is helpful in judging the maturity of the data and for visualizing the probable effect of additional follow-up.

1.9. NUMBER OF DEATHS

In the absence of censoring, the ith individual in a sample of size n has survival time T_i; $i = 1,2,\ldots,n$. For each subject, the potential censoring time C is defined as the period of observation such that observation on that individual ceases if death/failure has not occurred by then. In other words, C is the maximum observable time; in a medical follow-up study, this is the time period from enrollment to the conclusion of the study (date of data analysis). Also let $R(t)$ and $g(t)$ be the survival and probability density functions associated with C, viewed as a random variable,

$$R(t) = \Pr(C > t)$$

$$g(t)\,dt = \Pr(t \leq C \leq t + dt)$$

From a sample of size n, we have $\{C_i\}_{i=1}^n$, and $\{T_i\}_{i=1}^n$, the duration time t_i and status δ_i are defined as

$$t_i = \min(T_i, C_i)$$
$$= \delta_i T_i + (1 - \delta_i) C_i$$

$$\delta_i = \begin{cases} 0 & \text{if } C_i < T_i \text{ or } t_i = C_i \text{ (censored)} \\ 1 & \text{if } T_i < C_i \text{ or } t_i = T_i \text{ (event, i.e., uncensored)} \end{cases}$$

The previously mentioned crucial assumption on censoring that survival condition and reason of loss/censoring are independent indicates that the C_i's are stochastically independent of each other and of the true survival times T_i's. In a clinical trial with an enrollment period $(0, \pi_1)$ and a follow-up period (π_1, π_2), it is often assumed that the random variable C is distributed as uniform on the interval $(\pi_2 - \pi_1, \pi_2)$. Under this assumption, the proportion of subjects having events is given by

$$p = \Pr(\delta = 1)$$
$$= \int_0^{\pi_2} R(t)f(t)\,dt$$

For example, if T follows an exponential distribution with hazard ρ and C is distributed as uniform on $(\pi_2 - \pi_1, \pi_2)$, then

$$p = \int_0^{\pi_2 - \pi_1} \rho e^{-\rho t}\,dt + \int_{\pi_2 - \pi_1}^{\pi_2} \left(-\frac{t}{\pi_1} + \frac{\pi_2}{\pi_1}\right) \rho e^{-\rho t}\,dt$$

Since

(1) $$\int_0^{\pi_2 - \pi_1} \rho e^{-\rho t}\,dt = 1 - e^{-\rho(\pi_2 - \pi_1)}$$

and

(2) $$\int_{\pi_2 - \pi_1}^{\pi_2} \left(-\frac{t}{\pi_1} + \frac{\pi_2}{\pi_1}\right) \rho e^{-\rho t}\,dt$$

$$= -\frac{\rho}{\pi_1} \int_{\pi_2 - \pi_1}^{\pi_2} t e^{-\rho t}\,dt + \frac{\rho \pi_2}{\pi_1} \int_{\pi_2 - \pi_1}^{\pi_2} e^{-\rho t}\,dt$$

$$= \frac{\pi_2}{\pi_1}(e^{-\rho(\pi_2 - \pi_1)} - e^{-\rho \pi_2})$$

$$+ \frac{1}{\pi_1}\left[-(\pi_2 - \pi_1)e^{-\rho(\pi_2 - \pi_1)} + \pi_2 e^{-\rho \pi_2}\right.$$

$$\left. - \frac{1}{\rho}(e^{-\rho(\pi_2 - \pi_1)} - e^{-\rho \pi_2})\right]$$

$$= e^{-\rho(\pi_2 - \pi_1)} - \frac{1}{\rho \pi_1}(e^{-\rho(\pi_2 - \pi_1)} - e^{-\rho \pi_2})$$

therefore the proportion of uncensored cases is given by

$$p = 1 - \frac{1}{\rho \pi_1}(e^{-\rho(\pi_2 - \pi_1)} - e^{-\rho \pi_2})$$

and in the special case of no follow-up, $\pi_1 = \pi_2 = \pi$,

$$p = 1 - \frac{1}{\rho \pi}(1 - e^{-\rho \pi})$$

The expected number of deaths is np where n is the total enrollment (number of patients); this expected number of deaths—not n—would tell us whether we have a small study or a large study.

1.10. COMPUTATIONAL IMPLEMENTATION

Much of this book is concerned with arithmetic/algebraic procedures for the analysis of survival data. In many biomedical investigations, particularly those involving large quantities of data, the analysis—especially regression analysis—gives rise to difficulties in computational implementation. In these investigations it will be necessary to use statistical software specially designed to do these jobs. Most of the calculations described in this book can be readily carried out using statistical packages, and any student or practitioner of survival analysis will find the use of such packages essential. Two large packages that are widely used are: BMDP (two modules are designed especially for survival analysis: BMDP1L and BMDP2L) and SAS (Proc LIFETEST and Proc PHREG). The investigator contemplating the use of one of these commercial programs must read the specifications of the program before choosing options needed or suitable for any particular arithmetic/algebraic procedure. SAS program instructions are included in our examples where they were used.

Chapter 2 introduces some of the most widely used methods for estimating the survival function and related parameters. Chapter 3 is devoted to techniques for comparing survival distributions with applications to the analysis of epidemiological cohort studies and to the sample size determination. Chapter 4 deals with the identification of prognostic or risk factors related to survival time, covering standard procedures as well as providing mathematical foundations for some methods presented in Chapter 3; emphasis is on nonparametric procedures originating from Cox's proportional hazards multiple regression model.

EXERCISES

1.1. Given the following uncensored sample,

$$\{12, 7, 9, 4, 13\}$$

graph the empirical survival curve, calculate the area under this curve, and compare the result to the sample mean.

1.2. Consider a typical clinical trial with 3 years of enrollment and 2 years of follow-up. Assume that patients enter the study uniformly and have survival times exponentially distributed with an estimated median of 6 years. How many deaths are expected if the study enrolls 50 patients per year? Compute and compare the results from the following two designs:
   (a) Increase enrollment period by 1 year.
   (b) Increase follow-up period by 3 years.

1.3. A simple alternative to methods of survival analysis is the comparison of x-year survival rates. Consider this example:

Subject	Starting	Ending	Status (A/D)
1	01/80	01/90	A
2	06/80	07/88	D
3	11/80	10/84	D
4	08/81	02/88	D
5	04/82	01/90	A
6	06/83	11/85	D
7	10/85	01/90	A
8	02/86	06/88	D
9	04/86	12/88	D
10	11/86	07/89	D

Analysis date: 01/90; A = Alive, D = Dead.

For each subject, the potential censoring time is defined as the period of observation such that observation on that individual ceases if death/failure has not occurred by then (in a medical follow-up study, this is the time period from enrollment to the conclusion of the study).

(a) One way to define the 5-year survival rate is to use the proportion, among subjects with potential censoring times exceeding 5 years, whose failure time is observed to exceed 5 years. From the above given data, calculate the estimated 5-year survival rate, called p_1. (NOTE: Subjects whose potential censoring times are less than 5 years are not used in this calculation.)

(b) Suppose that subject 8 had actually entered on 02/84 but died at the same time (so that his survival time—as well as the group survival experience—would have been *improved*). Recalculate the estimated 5-year survival rate, called p_2. Compare p_2 to p_1 and comment on the result and the definition of this 5-year survival rate.

(c) Repeat (a) and (b) assuming that the analysis date is moved to 01/92 and ending date for subjects 1, 5, and 7 are also moved to 01/92.

1.4. Consider a distribution with hazard function

$$\lambda(t) = \theta e^t \quad \text{for} \quad t > 0 \quad \text{and} \quad \theta > 0$$

Find the survival function and median survival time.

1.5. Consider a distribution characterized by a hazard function that is proportional to $(\psi - t)^{\beta - 1}$ for $t < \psi$ and $0 < \beta < 1$.

(a) Show that the survival function can be expressed as

$$S(t) = \begin{cases} 1 & \text{if } t < 0 \\ \exp[-\theta\{\psi^\beta - (\psi - t)^\beta\}] & \text{for } 0 \leq t < \psi \\ 0 & \text{if } t \geq \psi \end{cases}$$

(note an abrupt mass extinction at $t = \psi$).

(b) Find the density $f(t)$.

(c) Find $S(t)$ and $f(t)$ for $\beta = 0$ (the generalized Pareto distribution) (Zelterman, 1992).

1.6. Koehler and McGovern (1990; also see Board, 1949; Goldman, 1984) presented a quasi-experimental study that was conducted to compare the relative effectiveness of two smoking cessation programs, an aversion-maintenance (AM) program and a nicotine fading-maintenance (NM) program. Subjects were recruited and invited to attend an information meeting at which they self-selected their preferred treatment. Both programs consisted of 17 sessions conducted over a 10-week period; each session lasted from 45 minutes to one hour and was led by a facilitator who was an ex-smoker. The authors denoted the time until a subject returns to smoking as the survival time, where a return to smoking was defined as the start of 10 days of regular smoking after an initial quit period of at least 24 hours. In this study the observed survival times are interval censored; for example, if a subject was observed as abstinent at 3 months but as a recidivist at 6 months, it is only known that his/her survival time is between 3 and 6 months. The authors then assumed a *limited failure population* (LFP, also called a *cure model*) where a proportion ρ of the population was assumed never fail (smokers who remain permanently abstinent after receiving treatment). The remaining $100(1-\rho)$ percent have their survival times following a Weibull distribution. Write down the formula for (combined) survival function for each program; the aim here is to compare the "cured fractions" ρ_{AM} versus ρ_{NM}. Note that this violates the basic mathematical property of a survival function because

$$\lim_{t \to \infty} S(t) \neq 0$$

but the model has also been used in other medical contexts as well, for example, patient survival from bone marrow transplant.

1.7. Suppose that survival time T of an individual is exponential with hazard ρ but not every individual has the same hazard. Consider ρ as a nonnegative random variable and assume that across populations, ρ is distributed as gamma with parameters (α, β). Find the marginal density, hazard, and survival functions for T (the Pareto distribution). We can use the idea to model survival from organ transplant using a Weibull distribution with a fixed shape parameter and gamma-distributed scale parameter. The limited failure population model of Exercise 6 can also be generalized with a gamma-distributed cured fraction ρ.

1.8. Find the survival function of the following piecewise exponential distribution:

$$\lambda(t) = \begin{cases} \lambda_1 & \text{if } 0 \leq t < \pi_1 \\ \lambda_2 & \text{if } \pi_1 \leq t < \pi_2 \\ \lambda_3 & \text{if } \pi_2 \leq t \end{cases}$$

where π_1, π_2 are known constants.

EXERCISES

1.9. Consider the following idea as a possible model for organ transplant. The survival time T of an individual from transplant depends on an unobservable transition time X during which the new organ may be rejected. Assume that the distribution of T given X is piecewise exponential with conditional hazard

$$\lambda(t \mid x) = \begin{cases} \lambda_1 & \text{if } t < x \\ \lambda_2 & \text{if } t \geq x \end{cases}$$

(it may be better fitted with piecewise Weibull, but piecewise exponential is assumed here for simplicity). The transition time varies from individual to individual (of course, not every individual will survive past transition time).

(a) Find the conditional survival function $S(t \mid x)$.
(b) Suppose that across the population X is distributed with density $f(x)$. Calculate the proportion of individuals surviving past their transition ages, that is,

$$\Pr(T \geq X)$$

(c) Find the marginal survival function of T if $f(x)$ is a gamma density [with parameters (α, β)].

One of the basic modeling ideas here is that medications used during an operation such as immune-suppression drugs (e.g., ALG) may affect only the transition time, but not survival after that period (Zelterman et al., 1994).

1.10. Suppose that data are available on a reasonably homogeneous group of patients with renal failure. All patients are initially on hemodialysis and the time at which they start this treatment is the time origin for each patient. All patients are observed until death. Depending on the availability of suitable donor kidneys, some patients in due course receive a kidney transplant. It is required to evaluate the benefit of having a transplant. Criticize qualitatively the following two procedures:
(a) Form two groups of patients, those never transplanted and those receiving a transplant. Compare the two distributions of time from entry to death, regardless of the time of transplant.
(b) Same as (a) except that for transplanted patients take a new time, origin at the instant of transplant.

Any recommendations of your own?

1.11. Construct an estimated survival curve for the data of Example 2 and verify that the area under this curve is equal to the sample mean.

1.12. Consider the remission times in Example 3.
  (a) Construct an estimated survival curve.
  (b) Assuming that these times are exponentially distributed with parameter hazard ρ, compute $\hat{\rho}$ and plot the survival function

$$\hat{S}(t) = \exp(-\hat{\rho}t)$$

in the same graph with the step function obtained in (a).

1.13. Repeat Exercise 1.12 with the data from Example 4.

1.14. Construct and plot two survival curves in the same graph, one for AG-positive patients and one for AG-negative patients of Example 5.

1.15. Referring to data of Example 5, use the Wilcoxon rank-sum test to compare survival times of AG-positive patients versus those of AG-negative patients.

1.16. Divide the sample in Example 6 into two groups:

Group 1: hospitalized patients ($x_1 \leq 60$)
Group 2: patients able to care for self ($x_1 = 70$–90)

Compare the survival times of these two groups using the Siegel–Tukey test.

1.17. Repeat Exercise 1.16 by applying the Wilcoxon test to deviations from the group means.

1.18. Consider the survival times of AG-negative patients in Example 5 and divide this sample into three groups.

Group 1: WBC $\geq 50{,}000$
Group 2: $10{,}000 \leq$ WBC $< 50{,}000$
Group 3: WBC $< 10{,}000$

Compare the survival times of these three groups using the Kruskal–Wallis test.

EXERCISES

1.19. Using the three groups of patients in Exercise 1.18, test for the trend (that higher WBC is associated with poorer survival) using the Jonckheere and Le tests.

1.20. Referring to the data of Example 6, investigate the relationship between survival time and time from diagnosis to randomization by calculating both Spearman's ρ and Kendall's τ.

CHAPTER 2

Estimation of Functions and Parameters

2.1. Estimation of the Survival Function
   Kaplan–Meier Curve
   Actuarial Method
   Mean, Median, and Other Parameters
   Standardization of Survival Rates
   Current Population Life Tables
2.2. Estimation of Parameters in Survival Models
   Some Survival Models
   Likelihood Function
   Exponential Model
   Weibull Model
   Special Use of the Standardized Mortality Ratio
2.3. Cohort Studies
   Method
   Model
   Results
   Healthy Worker Effects
2.4. Estimation of the Cumulative Hazard
2.5. Rank Correlation for Bivariate Censored Data
Exercises

Data analysis often depends on identifying a probability model that gives rise to observed data. In Section 2.2, a variety of simple probability models are considered; each model is a family of distributions specified by one or several parameters. Maximum-likelihood method will be applied for parameter estimation, in the presence of censoring. For the special case of survival analysis, however, statistical focus has shifted more and more to nonparametric approaches. While this book attempts to cover both the parametric and nonparametric methods, the emphasis is on nonparametric methods, which are more often used in health research. Section 2.1 concerns the estimation of the survival curve and an application, and Section 2.2 is involved with the estimation of parameters in parametric survival methods. Section 2.3 presents a health application, the analysis of epidemiological cohort studies; and Section 2.4 is devoted to the estimation of the cumulative hazard function. Finally, Section 2.5 describes a generalization of Kendall's tau for use with censored data.

2.1. ESTIMATION OF THE SURVIVAL FUNCTION

In this section we will discuss the estimation of the survival function and other related parameters by nonparametric or distribution-free methods. These methods are easy to understand and apply; and even if the main objective is to find a suitable model for the data, estimates and graphs obtained by nonparametric methods can be very helpful in choosing such a distribution.

As earlier indicated, what is special about the analysis of survival data is that at the time of data analysis, there are still subjects without the event. These subjects contain only partial information, for example, having survived 10 years or more, and are called *censored* cases. Example 12 of Chapter 1 is given here again for illustration:

Example 1. The remission times of 42 patients with acute leukemia were reported from a clinical trial undertaken to assess the ability of a drug called 6–mercaptopurine (6–MP) to maintain remission. Each patient was randomized to receive either 6–MP or placebo. The study was terminated after one year; patients have different follow-up times because they were enrolled sequentially at different times. Times in weeks were:

6–MP group: 6, 6, 6, 7, 10, 13, 16, 22, 23, 6^+, 9^+, 10^+, 11^+, 17^+, 19^+, 20^+, 25^+, 32^+, 32^+, 34^+, 35^+
Placebo group: 1, 1, 2, 2, 3, 4, 4, 5, 5, 8, 8, 8, 8, 11, 11, 12, 12, 15, 17, 22, 23

ESTIMATION OF THE SURVIVAL FUNCTION

in which a t^+ denotes a censored observation, that is, the case was censored after t weeks without a relapse. For example, "10$^+$" is a case enrolled 10 weeks before study termination and still remission-free at termination.

In the analysis of survival data, an important statistic that is of considerable interest is the "t-year survival rate" (in the context of the above sample, it is the t-week survival rate). The t-year survival rate is defined as

$$S(t) = \frac{\text{Number of individuals who survive longer than } t \text{ years}}{\text{Total number of individuals in the data set}}$$

As a function of time t, $S(t)$ is called the *survival function*

$$S(t) = \Pr[T > t]$$

The 5-year survival rate is commonly used in cancer research as a measure of a treatment effectiveness. In clinical trials, the differences between survival rates are indications of a treatment effect. Without censoring, the calculation of survival rates is straightforward, as seen in the previous chapter; for example, in the above placebo group,

$$\text{10-week survival rate} = \tfrac{8}{21} \times 100\%$$

$$= 38.1\%$$

because 8 of 21 cases survived 10 weeks or more.

In the following sections we will discuss methods appropriate for censored data; these methods are constructed so as to reduce to the above familiar empirical rates when data are not censored.

Kaplan–Meier Curve

We first introduce the *product-limit* (PL) method of estimating the survival rates; this is also called the *Kaplan–Meier* method (Kaplan and Meier, 1958).

Let

$$t_1 < t_2 < \cdots < t_k$$

be the distinct observed death times in a sample of size n from a homogeneous population with survival function $S(t)$ to be estimated ($k \leq n$). Let n_i be the number of subjects at risk at a time just prior to t_i ($1 \leq i \leq k$; these are cases whose duration time is at least t_i), and d_i the number of deaths at t_i. Consider $S(t)$ as a discrete function with probability mass at each t_i; we have

$$f(t_i) = \lambda_i \prod_{j=1}^{i-1} (1 - \lambda_j)$$

$$S(t_i) = \prod_{j=1}^{i} (1 - \lambda_j)$$

The conditional probability that d_i of the n_i subjects at risk die is given by

$$L_i = \frac{[f(t_i)]^{d_i} [S(t_i)]^{n_i - d_i}}{[S(t_i^-)]^{n_i}}$$

$$= \lambda_i^{d_i} (1 - \lambda_i)^{n_i - d_i}$$

where t_i^- is a time just before t_i. This is similar to a binomial process with n_i trials, d_i successes and probability of success λ_i. Given the data, the products

$$L = \prod L_i$$

$$= \prod_{i=1}^{k} \lambda_i^{d_i} (1 - \lambda_i)^{n_i - d_i}$$

can be viewed as a likelihood function on the space of survival functions $S(t)$; the maximum-likelihood estimate is the survival function $\hat{S}(t)$ that maximizes L. The results are

$$\hat{\lambda}_i = d_i / n_i$$

ESTIMATION OF THE SURVIVAL FUNCTION

and

$$\hat{S}(t) = \prod_{t_i \le t} \left(1 - \frac{d_i}{n_i}\right)$$

which is called the *product-limit estimator* or *Kaplan–Meier's estimator* for $S(t)$. Standard-likelihood method and error propagation also yield

$$\widehat{\text{Var}}[\log \hat{S}(t)] = \sum_{t_i \le t} (1 - \hat{\lambda}_i)^2 \widehat{\text{Var}}(1 - \hat{\lambda}_i)$$

$$= \sum_{t_i \le t} \frac{d_i}{n_i(n_i - d_i)}$$

leading to an expression for the asymptotic variance of $\hat{S}(t)$:

$$\widehat{\text{Var}}[\hat{S}(t)] = \hat{S}^2(t) \sum_{t_i \le t} \frac{d_i}{n_i(n_i - d_i)}$$

which is known as Greenwood's formula (Greenwood, 1926). A direct application of this formula to small and large values of t may result in 95 percent confidence limits outside the interval $[0,1]$. To avoid these impossible values, as well as to improve normal approximations, 95 percent confidence intervals may be first formed on the log scale followed by exponentiating both resulting confidence limits to obtain

$$\hat{S}(t) \exp[\pm 1.96 \hat{s}(t)]$$

where

$$\hat{s}^2(t) = \sum_{t_i \le t} \frac{d_i}{n_i(n_i - d_i)}$$

According to the product-limit method, survival rates for the 6–MP group are calculated by constructing a table such as Table 2.1, with five columns as shown.

TABLE 2.1. Estimation of Survival Rates for 6–MP Group of Example 1

(1) t_i	(2) n_i	(3) d_i	(4) $1 - \dfrac{d_i}{n_i}$	(5) $\hat{S}(t_i)$
6	21	3	.8571	.8571
7	17	1	.9412	.8067
10	15	1	.9333	.7529
13	12	1	.9167	.6902
16	11	1	.9091	.6275
22	7	1	.8571	.5378
23	6	1	.8333	.4482

Note: An SAS program would include these instructions:

PROC LIFETEST METHOD=KM;
TIME WEEKS*RELAPSE(0);

where WEEKS is the variable name for duration time, RELAPSE the variable name for survival status, "0" is the coding for censoring, and KM stands for Kaplan–Meier method.

To obtain $\hat{S}(t)$, multiply all values in column 4 up to and including t. From Table 2.1, we have, for example:

6-week survival rate is 85.71%

22-week survival rate is 53.78%

We have, for example, using the log transformation:

$$\text{SE}[\hat{S}(7)] = (.8067)\left\{\left[\frac{3}{(21)(18)} + \frac{1}{(17)(16)}\right]\right\}^{1/2}$$

$$= .0869$$

and a 95 percent confidence interval for $S(7)$ is (.6804, .9565) using the method previously mentioned, going through the logarithmic transformation.

A convenient graphical way of displaying survival rates is by means of a *survival curve* or *Kaplan–Meier curve*. A survival curve

ESTIMATION OF THE SURVIVAL FUNCTION

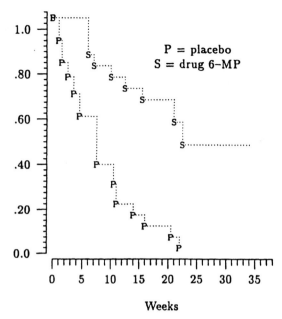

FIGURE 2.1. Survival rates for data in Example 1.

is a graph in which

1. the horizontal scale represents the times marked at uncensored observations and
2. the vertical scale represents the survival rates

so that each uncensored observation is represented by a dot, and these dots are joined so as to form a step curve. For the data of Example 1, survival rates for the two groups are displayed in the same graph as shown in Figure 2.1.

Actuarial Method

The Kaplan–Meier method is applicable for any survival data set; however, for a large data set—say 100 or more subjects—it may be much more convenient to group the times into intervals. The process is similar to the formation of a frequency table and a histogram in elementary statistics, and the method is referred to as the *actuarial method*. A concrete example, Example 2, is adapted to illustrate this method; it consists of eight columns, following these outlines:

1. The choice of time intervals will depend on the nature of the data and the size of the data set, and guidelines for the formation of a frequency table are equally applicable here.
2. **and 3.** The patients are classified according to the time interval during which their condition was last recorded. If the report was a death, the patient is counted in column (2), and if censored, the patient is counted in column (3). Time for each patient is calculated from his or her enrollment in the study.
3. The number of patients living at the start of the intervals is obtained by cumulating columns (2) and (3) from the foot. For example, the number alive at 10 years is $21 + 26 = 47$, and the number alive at 9 years is $47 + 2 + 5 = 54$, and so on.

Example 2. Survival of patients with a particular form of malignant disease:

(1)	(2)	(3)	(4)	(5)	(6)	(7)	(8)
Time (years) t to $(t+1)$	Last reported Died d	Last reported Censored c	Living at the start n	Number at risk n'	Interval death rate q	Interval survival rate p	Survival rate at t (%) $S(t)$
0–1	90	0	374	374.0	.2406	.7594	100.0
1–2	76	0	284	284.0	.2676	.7324	75.9
2–3	51	0	208	208.0	.2452	.7548	55.6
3–4	25	12	157	151.0	.1656	.8344	42.0
4–5	20	5	120	117.5	.1702	.8298	35.0
5–6	7	9	95	90.5	.0773	.9227	29.1
6–7	4	9	79	74.5	.0537	.9463	26.8
7–8	1	3	66	64.5	.0155	.9845	25.4
8–9	3	5	62	59.5	.0504	.9496	25.0
9–10	2	5	54	51.5	.0388	.9612	23.7
10+	21	26	47	—	—	—	22.8

Note: If input consist of raw data, an SAS program would include these instructions:

PROC LIFETEST METHOD=ACT INTERVALS=1,2,3,4,5,6,7,8,9,10;
TIME YEAR*FAILURE(0);

where YEAR is the variable name for duration time, FAILURE the variable name for survival status, "0" is the coding for censoring, ACT stands for actuarial method, and numbers listed are cutpoints needed for dividing time scale into intervals.

ESTIMATION OF THE SURVIVAL FUNCTION

5. The number of patients at risk during interval t to $(t+1)$ is

$$n' = n - c/2$$

The purpose of this formula is to provide a denominator for the next column. The adjustment from n to n' is needed because the c censored subjects are necessarily at risk for only part of the interval. The basic assumption here is that, on the average, these subjects are at risk for half of the interval.

6. The interval death rate is

$$q = d/n'$$

(conditional on survival at the beginning of the interval). For example, during the first interval

$$q = \frac{90}{374.0}$$
$$= .2406$$

7. The interval survival rate is

$$p = 1 - q$$

8. To obtain $S(t)$, the survival rate at t years, multiply all values in column (7) up to and including that for interval t to $(t+1)$. For example,

$$\text{One-year survival rate} = 75.9\%$$
$$\text{Five-year survival rate} = 29.1\%$$

Survival rate calculated by the actuarial method can also be displayed by the survival curve similar to that for rates obtained by the Kaplan–Meier method. The only difference here is that the dots in the diagram are usually connected by straight lines as shown in Figure 2.2.

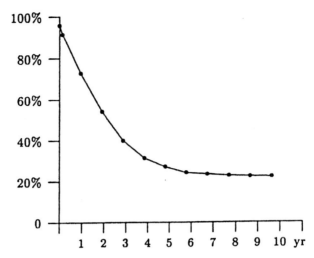

FIGURE 2.2. Survival curve for cancer patients of Example 2.

Finally, it is noted that this approximation—the actuarial method—is still useful even if the user has access to computer packages for constructing the Kaplan–Meier curve because it provides an estimate for the "hazard" or "risk" function, which is not easily obtained with the Kaplan–Meier method and which is useful and needed for certain more advanced analyses.

Mean, Median, and Other Parameters

The mean survival time μ can, as shown in Section 1.3, be represented by the area under the survival curve. Therefore, the area under an estimated curve—such as the one obtained by the product-limit method or by the life-table method—serves as an estimate, $\hat{\mu}$.

Example 3. It has been expected that life expectancy of people with mental retardation is shorter than that of the general population. However, exact estimates have not been available. In order to obtain these, data was collected for 99,543 persons with developmental disabilities, including mental retardation, who received services from the California Department of Developmental Services between March 1984 and October 1987. Subjects were divided into three groups and the following table is for the most se-

ESTIMATION OF THE SURVIVAL FUNCTION

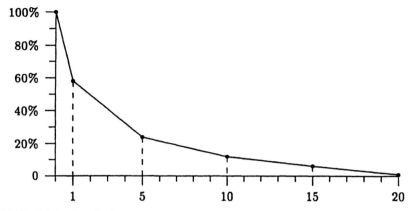

FIGURE 2.3. Survival curve for patients with mental retardation of Example 3.

vere cases ($n = 1550$); these people had severe deficits in cognitive function, were incontinent and immobile, and required tube feeding.

Age Interval	Survival Rate (%) at Beginning
0–1	100
1–4	59
5–9	21
10–14	7
15–19	3
20–24	1

The survival curve is shown in Figure 2.3, and the mean survival time is approximately

$$\hat{\mu} = (1)(1.00 + .59)/2 + (4)(.59 + .21)/2 + (5)(.21 + .07)/2$$
$$+ (5)(.07 + .03)/2 + (5)(.03 + .01)/2 + (5)(.01 + 0)/2$$
$$\cong 3.48$$

In this calculation,

$$(1)(1.00 + .59)/2$$

is the area of the first trapezoid and

$$(4)(.59 + .21)/2$$

is the area of the second trapezoid, and so forth. The sum of the areas of these trapezoids is the area under the estimated survival curve.

It can be seen from the method and Example 3 that the mean survival time is estimable if and only if we have a *complete* curve, that is, the curve drops to zero (the 1 percent at 20 years is considered negligible in this example). It is obvious that this requirement is impractical for most medical follow-ups (see Examples 1 and 2), and therefore the mean is not a popular measure in survival analysis.

In many studies, instead of the mean, it is easier to obtain other measures of location such as the *median*, $\hat{T}_{.5}$, and other quantiles (e.g., 25th and 75th percentiles). For Example 3, we have

$$\text{1-year survival rate} = 59\%$$
$$\text{5-year survival rate} = 21\%$$

a linear interpolation gives

$$\hat{T}_{.5} = 1 + (4)\frac{.59 - .50}{.59 - .21}$$
$$= 1.95$$

or a median survival time of about 2 years. This figure and estimates for other quantiles and their standard errors can be easily obtained using computer packages.

Standardization of Survival Rates

Survival rates, as estimated by the product-limit method, can be used for population description and may be suitable for investigation of variations between subpopulations or over time. However, the comparisons of these rates are often invalid because the populations to be

ESTIMATION OF THE SURVIVAL FUNCTION

compared may be different with respect to one or more characteristics, called *confounders*. To overcome this difficulty, an adjusted rate is needed; the adjustment removes the difference in the composition with respect to a confounder.

Example 4. End-stage renal disease is an irreversible chronic renal failure. End-stage renal disease (ESRD) may be caused by progressive primary renal disease, kidney damage due to other diseases, or acute irreversible damage resulting in permanent renal failure. For analytical purposes, these causes are classified into eight disease categories:

Diabetes
Hypertension (causing nephrosclerosis)
Glomerulonephritis
Cystic kidney disease
Other urologic disease (primarily obstructive nephropathy)
Other identifiable diseases
Cause unknown
Missing information on the cause

In Table 2.2 we consider two "incident cohorts" of patients. The first cohort consists of all patients receiving their first ESRD therapy in the year 1980, and the second cohort, in the year 1990. Table 2.2 shows that the 1990 cohort has lower one-year survival rate (78.21 vs. 80.35 percent for the 1980 cohort). However, a closer examination shows

(i) The percent of patients caused by diabetes in 1990 is much higher than in 1980 (34.63 vs. 12.50 percent).
(ii) Patients with diabetes have the lowest survival—and this is true throughout the 1980s.

These observations make it essential to adjust the survival rates of the two cohorts in order to make a valid comparison.

A simple way to adjust crude rates, called the *direct method*, is to choose a standard group and to apply diagnosis-specific rates obtained from the two cohorts under investigation. It is important to note that any population could be chosen as "standard" because

TABLE 2.2. Kaplan–Meier One-Year Survival Rates (%) for the 1980 and 1990 Incident Cohorts

	1980 Cohort		1990 Cohort	
Primary Diagnosis	No. of Patients	One-Year Survival Rate	No. of Patients	One-Year Survival Rate
Diabetes	2,135	72.42%	14,597	75.08%
Hypertension	2,431	81.51	11,699	75.97
Glomerulonephritis	2,077	89.81	5,338	86.26
Cystic kidney	604	93.13	1,300	92.23
Other urologic	1,074	85.13	2,071	79.18
Other cause	394	82.30	2,444	74.76
Unknown cause	1,404	82.83	2,058	78.71
Cause missing	6,962	77.04	2,642	83.66
Total (cohort)	17,081	80.35	42,149	78.21

adjusted rates have real meaning only as relative comparisons; they need not reflect data from an actual population. For example, we can choose the 1980 cohort as the standard population and adjust the rate for the 1990 cohort. The procedure consists of the following steps.

(i) The "expected" numbers of survivors in the standard population are computed for each diagnosis category. For example, the size of the diabetic group in the standard population (i.e., 1980 cohort) was 2135 and the 1990 one-year survival rate for this group was 75.08 percent. Therefore the expected number of survivors is

$$(2135)(.7508) \cong 1,602.96$$

(ii) The expected numbers of survivors for all categories are added up to obtain the total number of expected survivors and the adjusted one-year survival rate is calculated:

$$\text{Adjusted rate} = \frac{\text{Total number of expected survivors}}{\text{Standard population size}}$$

Detailed calculations are shown in Table 2.3.

ESTIMATION OF THE SURVIVAL FUNCTION

TABLE 2.3. Direct Method for Rate Standardization

Primary Diagnosis	1980 Patients	One-year 1990 Survival Rate	Expected Survivors
Diabetes	2,135	.7508	1,602.96
Hypertension	2,431	.7597	1,846.83
Glomerulonephritis	2,077	.8626	1,791.62
Cystic kidney	604	.9223	557.07
Other urologic	1,074	.7918	850.39
Other cause	394	.7476	294.55
Unknown cause	1,404	.7871	1,105.09
Cause missing	6,962	.8366	5,824.41
Totals	17,081		13,872.92

The one-year adjusted survival rate for the 1990 cohort is

$$\frac{13{,}872.92}{17{,}081} \times 100 = 81.22\%$$

which is slightly higher than the observed rate for the 1980 cohort of 80.35 percent. Of course, in this adjustment only cause differentials are removed; other factors, such as age, race, etc. should also be considered.

As seen earlier, when we use the product-limit method, variance of the estimated rate can be obtained using Greenwood's formula. The above adjusted rate takes the form

$$r = \sum w_x r_x$$

with r_x being the group-specific estimated rate and w_x the fraction of the standard population belonging to category x. It follows that variance of the adjusted rate r may be expressed as

$$\text{Var}(r) = \sum w_x^2 \text{Var}(r_x)$$

For the 1990 ESRD cohort, standard errors for course-specific rates are given in Table 2.4. The adjusted rate is .8122 with a cor-

TABLE 2.4. Standard Errors for 1990 Cohort Rates

		1990 Cohort	
Primary Diagnosis	1980 Patients	One-Year Survival Rate, r_x	Standard Error, s_x
Diabetes	2,135	.7508	.0037
Hypertension	2,431	.7597	.0040
Glomerulonephritis	2,077	.8626	.0048
Cystic kidney	604	.9223	.0074
Other urologic	1,074	.7918	.0091
Other cause	394	.7476	.0089
Unknown cause	1,404	.7871	.0091
Cause missing	6,962	.8366	.0072
Cohort	17,081	.7821	.0020

responding variance of

$$s^2 = \sum w_x^2 s_x^2$$
$$= (.0032)^2$$

or a standard error of .0032; the observed rate and its standard error were .7821 and .0020, respectively.

Current Population Life Tables

There are two types of life tables: *cohort* (or *follow-up*) life tables and the *current* life table.

The cohort or follow-up life table, as seen from previous sections, provides a longitudinal perspective in that it follows the mortality experience of a particular cohort over time. The current life table is just a way to summarize death rates for a specific period and may be characterized as "cross sectional." Unlike the cohort life table, the current life table does not represent the mortality experience of any group. Rather, the current life table considers a hypothetical cohort, of size 100,000 say, and assumes that it is subject to the age-specific death rates observed for an actual population during a particular period. Thus, for example, the U.S. current life table for 1987 assumes a

TABLE 2.5. Abridged U.S. Life Table, 1987, for White Males

Age Interval	No. Living at Beginning of Age Interval
0–1	100,000
1–5	99,037
5–10	98,834
10–15	98,698
15–20	98,539
20–25	97,970
25–30	97,204
30–35	96,448
35–40	95,584
40–45	94,539
45–50	93,187
50–55	91,136
55-60	87,834
60–65	82,710
65–70	75,095
70–75	65,101
75–80	51,843
80–85	36,587
85 and over	21,208

hypothetical cohort subject throughout its lifetime to the age-specific death rates prevailing for the actual American population in 1987. Tables published by the National Center for Health Statistics may be complete (containing life table values for single years of age) or abridged (by 5-year age groups). The complete tables are useful in the analyses of standardized mortality ratios in occupational health. Table 2.5 provides such a sample, an abridged table for white males, 1987.

2.2. ESTIMATION OF PARAMETERS IN SURVIVAL MODELS

Even though our main focus concerns the relationship between survival time and explanatory variables (or covariates), it is necessary to consider briefly survival time distributions and the estimation of parameters in these models. Throughout the literature on survival analysis, certain models have been used repeatedly; for example, exponential and Weibull models. Some others are gaining popularity; for example, log-logistic and Gompertz-Makeham models.

Some Survival Models

As before, T is a random variable representing survival time, and $t > 0$ represents a typical point in this range. $S(t)$ and $\lambda(t)$ represent the survival and hazard functions of T, respectively. The exponential distribution is obtained by taking the hazard function to be a constant λ over the range of T; the survival function is given by

$$S(t) = \exp(-\lambda t)$$

An important generalization of the exponential distribution allows for a *power* dependence of the hazard on time. This yields the two-parameter Weibull distribution with hazard function

$$\lambda(t) = \lambda p (\lambda t)^{p-1}$$

for $\lambda, p > 0$. This hazard is monotone (decreasing for $p < 1$ and increasing for $p > 1$) and reduces to the exponential (constant) hazard if $p = 1$; p is called the *shape parameter*.

We have the log-normal model when $Y = \log T$ is distributed as normal (Gaussian); this model is less convenient computationally because the survival and hazard functions do not have closed-form expressions. The hazard is not monotonic; it has value 0 at $t = 0$, increases to maximum and then decreases, approaching zero as t becomes large—applications include chronic leukemia and Hodgkin's disease. The log-logistic model is defined similarly with Y having a logistic distribution.

The Gompertz model is another generalization of the exponential distribution, which allows for an *exponential* dependence of the hazard on time,

$$\lambda(t) = \alpha \exp(\beta t)$$

ESTIMATION OF PARAMETERS IN SURVIVAL MODELS

and the Gompertz-Makeham model adds an initial fixed component to the hazard function

$$\lambda(t) = \lambda + \alpha \exp(\beta t)$$

We consider now a sample of survival data

$$\{(t_i, \delta_i)\}_{i=1}^n$$

where t_i is the duration and δ_i the survival status of the ith subject. Suppose that the survival model is specified up to one or several parameters θ with the survival function $S(t)$

$$S(t) = \Pr[T > t]$$

and density function $f(t)$

$$f(t)dt = \Pr[t \leq T \leq t + dt]$$

As mentioned previously, an important feature of survival analysis is the need to accommodate right censoring in the data. Consider the case of random censorship and assume that the potential censoring time C_i for the ith subject is a random variable with survival and density functions $R_i(t)$ and $g_i(t)$ (see Chapter 1), respectively ($i = 1, 2, \ldots, n$), and further assume that C_i's are stochastically independent of each other and of the survival times.

Likelihood Function

We now have

(i)
$$\Pr[t_i \in (t, t + dt), \delta_i = 1] = \Pr[T_i \in (t, t + dt), C_i > t]$$
$$= R_i(t)f(t)dt$$

(ii)

$$\Pr[t_i \in (t, t+dt), \delta_i = 0] = \Pr[T_i > t, C_i \in (t, t+dt)]$$
$$= S(t)g_i(t)\,dt$$

Therefore, given the data $\{(t_i, \delta_i)\}_{i=1}^n$, the likelihood function is (up to a multiple constant)

$$L = \prod_{i=1}^n [R_i(t_i)f(t_i)]^{\delta_i}[S(t_i)g_i(t_i)]^{1-\delta_i}$$

$$= \prod_{i=1}^n \{[R_i(t_i)]^{\delta_i}[g_i(t_i)]^{1-\delta_i}\}\{[f(t_i)]^{\delta_i}[S(t_i)]^{1-\delta_i}\}$$

$$= \prod_{i=1}^n [f(t_i)]^{\delta_i}[S(t_i)]^{1-\delta_i} \quad \text{(up to a multiple constant)}$$

$$= \prod_{i=1}^n [\lambda(t_i)]^{\delta_i} S(t_i)$$

because the term

$$[R_i(t_i)]^{\delta_i}[g_i(t_i)]^{1-\delta_i}$$

does not involve the parameters of the survival model (this model specifies the distribution of T, which is independent from C, i.e., noninformative censoring).

Exponential Model

$$L(\lambda) = \prod_{i=1}^n \lambda^{\delta_i} e^{-\lambda t_i}$$
$$= \lambda^{\sum \delta_i} e^{-\lambda \sum t_i}$$
$$= \lambda^d e^{-\lambda T}$$

ESTIMATION OF PARAMETERS IN SURVIVAL MODELS

where

$$d = \sum_{i=1}^{n} \delta_i$$

$$= \text{number of events}$$

$$T = \sum_{i=1}^{n} t_i$$

$$= \text{total duration time}$$

This leads to

$$\ln L = d \ln \lambda - \lambda T$$

so that from

$$0 = \frac{d}{d\lambda} \ln L$$

$$= \frac{d}{\lambda} - T$$

we have

$$\hat{\lambda} = \frac{d}{T}$$

the usual *rate* frequently quoted in clinical research: deaths per treatment month (or person-year, etc.).

From

$$-\frac{d^2}{d\lambda^2} \ln L = \frac{d}{\lambda^2}$$

the variance of $\hat{\lambda}$ is given by

$$\text{Var}(\hat{\lambda}) = \frac{\lambda^2}{E(d)}$$

There are three options for dealing with the denominator:

(i) Assume $\{C_i\}_{i=1}^n$ are independent and identically distributed with certain survival function $R(t)$ as in Chapter 1. This leads to

$$E(d) = nE(\delta_i)$$

$$= n \int_0^{\pi_2} R(t)f(t)\,dt$$

For example, if C is distributed as uniform on $(\pi_2 - \pi_1, \pi_2)$ in a clinical trial, then

$$E(d) = n\left\{1 - \frac{1}{\lambda \pi_1}[e^{-\lambda(\pi_2 - \pi_1)} - e^{-\lambda \pi_2}]\right\}$$

(see the last part of Chapter 1).

(ii) Treat the potential censoring times $\{C_i\}_{i=1}^n$ as fixed. The problem is for noncensored cases, C_i's may be unknown in designs other than clinical trials. If we treat the C_i's as fixed, then

$$E(d) = \sum e - \lambda^{C_i}$$

(iii) The easiest way is to use observed information so that

$$\operatorname{Var}(\hat{\lambda}) \doteq \hat{\lambda}^2 / d$$

$$= d/T^2$$

The 95% confidence interval for λ, say, is given by

$$\frac{d \pm 1.96\sqrt{d}}{T}$$

Weibull Model

We have

$$L = \prod_{i=1}^n [\lambda p(\lambda t_i)^{p-1}]^{\delta_i} \exp[-(\lambda t_i)^p]$$

ESTIMATION OF PARAMETERS IN SURVIVAL MODELS

or

$$\ln L = \sum_{i=1}^{n} \{\delta_i[\ln p + p\ln \lambda + (p-1)\ln t_i] - (\lambda t_i)^p\}$$

The maximum-likelihood estimates $\hat{\lambda}$ and $\hat{p}$ are solutions of:

$$\frac{\partial}{\partial \lambda} \ln L = 0$$

$$\frac{\partial}{\partial p} \ln L = 0$$

In this case, there are no closed-form solutions; however, iterative numerical results may be obtained using a method such as Newton–Raphson (Cohen, 1965).

Example 5. Recall the data set on leukemia patients of Example 1 where we have remission times in weeks:

6–MP group: 6, 6, 6, 7, 10, 13, 16, 22, 23, 6+, 9+, 10+, 11+, 17+, 19+, 20+, 25+, 32+, 32+, 34+, 35+

Placebo group: 1, 1, 2, 2, 3, 4, 4, 5, 5, 8, 8, 8, 8, 11, 11, 12, 12, 15, 17, 22, 23

(i) If exponential models are assumed, we have:
For 6–MP group,

$$d_1 = 9 \qquad T_1 = 359$$

leading to

$$\hat{\lambda}_1 = .0251 \qquad SE(\hat{\lambda}_1) = .0084$$

For placebo group,

$$d_2 = 21 \qquad T_2 = 182$$

leading to

$$\hat{\lambda}_2 = .1154 \qquad SE(\hat{\lambda}_2) = .0252$$

(ii) Fitting Weibull models to the data of these two groups, we obtain the following results:
For 6–MP group,

$$\hat{\lambda}_1 = .0296 \qquad \hat{p}_1 = 1.3537$$

For placebo group,

$$\hat{\lambda}_2 = .1055 \qquad \hat{p}_2 = 1.3705$$

indicating a drug effect on the scale parameter but not on the shape parameter.

It is also interesting to note that for Weibull models, the hazards are proportional if and only if the two distributions have the same shape parameter p; in this case,

$$\frac{\lambda_1(t)}{\lambda_2(t)} = \frac{\lambda_1 p(\lambda_1 t)^{p-1}}{\lambda_2 p(\lambda_2 t)^{p-1}}$$
$$= \lambda_1^p / \lambda_2^p$$

Special Use of the Standardized Mortality Ratio

The United States Renal Data System (USRDS) is a federal agency that pursues the collection and analysis of information on the incidence, prevalence, morbidity, and mortality of ESRD in the United States. Mortality tables of the USRDS allow description of the national mortality rates among prevalent dialysis patients in 5-year age groups and four major categories of causes of ESRD (diabetes, hypertension, glomerulonephritis, and others, excluding missing diagnosis), and race. Based on these tables, a method has been proposed (Wolfe et al., 1992) in order to compare local ESRD mortality rates to national rates. This methodology adjusts for patient age, race, and cause of ESRD and can be summarized in Example 6.

Example 6. Data from three most recent years are pooled together to serve as a national death rate reference. In Table 2.6, we

TABLE 2.6. ESRD Death Rates During 1987–1989 per 1000 Patient Years at Risk Among Patients Alive on January 1, by Age, Primary Disease, and Race

	Total dialysis patients (never Tx) on January 1 (censored at first transplant)									
	Diagnosis									
Age as of Jan. 1	All (includes missing diag.)		Diabetes		Hypertension		Glomerulo-nephritis		Other (excludes missing diag.)	
	Black	White	Black	White	Black	White	Black	White	Black	White
0–14	60.6	48.0	—	—	—	—	—	—	—	—
15–19	52.7	40.5	—	—	—	—	—	—	—	—
20–24	76.9	65.1	227.9	177.3	57.4	53.5	42.1	47.1	104.4	60.1
25–29	98.2	95.8	166.9	234.7	76.4	83.7	81.8	51.3	111.4	59.8
30–34	128.1	126.0	194.9	273.6	97.5	85.4	133.5	48.3	134.0	79.8
35–39	112.7	125.7	187.0	283.9	88.6	61.7	111.9	63.5	132.3	66.4
40–44	114.6	153.4	182.0	304.5	97.2	118.5	103.0	75.1	110.6	104.6
45–49	123.4	166.3	172.3	322.9	104.2	125.1	96.7	94.5	126.9	116.1
50–54	148.2	201.4	209.7	342.9	127.7	186.2	119.6	129.2	126.9	139.4
55–59	178.1	231.7	228.5	386.9	149.3	221.1	153.4	164.2	152.0	170.4
60–64	213.5	278.5	250.5	421.0	196.4	270.3	196.6	212.4	197.1	216.1
65–69	255.3	321.5	300.9	454.5	228.0	321.7	215.0	261.2	242.1	265.1
70–74	312.9	384.9	356.1	539.0	292.4	408.4	301.2	316.6	301.2	320.0
75–79	356.1	451.4	406.2	598.1	343.7	470.7	283.6	402.1	361.3	402.3
80–84	414.7	555.1	508.0	713.6	397.1	592.0	359.5	466.3	365.3	517.3
85+	507.4	670.7	579.7	755.8	467.0	689.7	638.2	648.7	560.7	609.7
Total	199.1	278.8	259.9	402.5	189.5	343.5	149.9	201.7	190.2	228.1

list the results for 1987–1989; rates were calculated assuming the exponential distribution and expressed as "death per 100 patient years at risk" (the missing cells were due to small cell sizes—very few patients were less than 20 years of age). Each rate is computed as the number of deaths divided by the sum of the lengths of follow-up in the group. To compare mortality in a study group to the national mortality, the number of years of dialysis therapy is first determined for each patient. Multiplying this number of follow-up years by the corresponding national death rate determined by the age, race, and diagnosis of the patient, the result is the "expected number of deaths." After these individual calculations, we

TABLE 2.7. Example for Using Mortality Tables (Table 2.6) to Compare Local and National ESRD Mortality Rates

| | | | | | | | National Rates | | | |
| | | | | | | | From Table | Standard per | | |
Patient	Age at Start	Race	Primary Disease	Start Date	Stop Date	Days at Risk	per 1000 Pat. Yrs	Pat. Day	Exp. Deaths	Obs. Deaths
1	76	White	Hyperten	12/31/87	8/4/88	217	470.7	0.00129	0.280	1
2	53	Black	Hyperten.	12/31/87	12/31/88	366	127.7	0.00035	0.128	0
3	29	White	Other	12/31/87	11/23/88	328	59.8	0.00016	0.054	1
4	37	White	Other	12/31/87	12/31/88	366	66.4	0.00018	0.067	0
5	23	Black	Diabetes	12/31/87	7/3/88	185	227.9	0.00062	0.116	0
6	65	Black	Hyperten.	12/31/87	8/6/88	219	228.0	0.00062	0.137	1
7	72	White	Glomer.	12/31/87	9/19/88	263	316.6	0.00087	0.228	1
8	70	Black	Glomer.	12/31/87	10/4/88	278	301.2	0.00083	0.229	1
9	83	White	Glomer.	12/31/87	11/19/88	324	466.3	0.00128	0.414	1
10	17	Black	—	12/31/87	12/28/88	363	52.7	0.00014	0.052	1
11	51	Black	Hyperten.	12/31/87	12/31/88	366	127.7	0.00035	0.128	0
12	54	White	Diabetes	12/31/87	12/31/88	366	342.9	0.00094	0.344	0
13	61	White	Diabetes	12/31/87	12/31/88	366	421.0	0.00115	0.422	0
14	78	White	Other	12/31/87	12/3/88	338	402.3	0.0011	0.373	1
15	47	White	Diabetes	12/31/87	12/31/88	366	322.9	0.00088	0.324	0
16	59	White	Diabetes	12/31/87	12/13/88	348	386.9	0.00106	0.369	0
17	66	White	Diabetes	12/31/87	11/24/88	329	454.5	0.00125	0.410	1
18	51	Black	Diabetes	12/31/87	12/31/88	366	209.7	0.00057	0.210	0
19	83	White	Hyperten.	12/31/87	10/17/88	291	592.0	0.00162	0.472	1
20	80	White	Glomer.	12/31/87	12/31/88	366	466.3	0.00128	0.468	0
Total									5.223	10

add up the expected deaths for all patients to obtain the total number of expected deaths (say, e); this sum is then compared to the observed number of deaths (say, d) for that study group, namely ratio

$$\text{SMR} = d/e$$

called the *standardized mortality ratio* (SMR). Table 2.7 gives an example of a small treatment center (20 patients). The standardized mortality ratio is

$$\text{SMR} = 10/5.223$$
$$= 1.91$$

showing that this treatment center has 91 percent higher mortality than is expected from the national mortality rate. Although the calculation was described for each individual patient, with very large groups or centers, it may be more efficient to sort the patients into age-race-diagnosis strata, calculating the total follow-up time for each stratum, multiplying this number by the corresponding national death rate, and summing up the results to obtain the number of expected deaths.

2.3. COHORT STUDIES

Cohort studies are designs in which one enrolls a group of healthy persons and follows them over certain periods of time; examples include occupational mortality studies. The cohort study design focuses attention on a particular exposure rather than a particular disease as in case-control studies. Advantages of a longitudinal approach include the opportunity for more accurate measurement of exposure history and a careful examination of the time relationships between exposure and disease.

Method

The observed mortality of the cohort under investigation often needs to be compared with that expected from the death rates of the national population (served as standard), with allowance made for age, sex, race, and time period. Rates may be calculated either for total deaths or for separate causes of interest. The statistical method is often referred to as the *person-years method*. The basis of this method is the comparison of the observed number of deaths, d, from the cohort with the mortality that would have been expected if the group had experienced death rates similar to those of the standard population of which the cohort is a part. The expected number of deaths is usually calculated using published national life tables and the method is similar to that of indirect standardization of death rates.

Each member of the cohort contributes to the calculation of the expected deaths for those years in which he or she was at risk of dying during the study. There are three types of subjects:

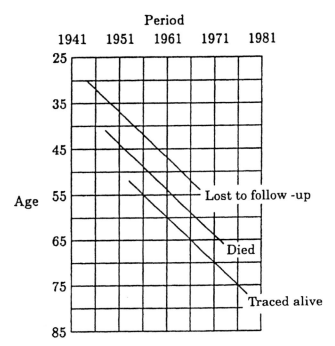

FIGURE 2.4. Representation of basis of subject-years method. Each subject is represented by a diagonal line that starts at the age and year at which the subject entered the study and continues as long as the subject is at risk of dying in the study.

(i) Subjects still alive on the analysis date
(ii) Subjects who died on a known date within the study period
(iii) Subjects who are lost to follow-up after a certain date. These cases are a potential source of bias; effort should be expended in reducing the number of subjects in this category.

Figure 2.4 shows the situation illustrated by one subject of each type.

In Figure 2.4, each subject is represented by a line starting from the year and age at entry and continuing until the study date, the date of death, or the date the subject was lost to follow-up. Period and age are divided into 5-year intervals corresponding to the usual availability of referenced death rates. Then a quantity, r, is defined for each individual as the cumulative risk over the follow-

COHORT STUDIES

up period:

$$r = \sum xw$$

where summation is over each square in Figure 2.4 entered by the follow-up line, x is the time spent in a square, and w is the corresponding death rate for the given age-period combination. For the cohort, the individual values of r are added to give the expected number of deaths:

$$e = \sum r$$

The quantity r is often referred to as the *expected mortality* for the individual. It is the cumulative risk, not a probability of dying; those subjects surviving beyond their expectation of life will have values of r greater than unity. However, for practical purposes, r may be approximated by the probability of dying, which can be calculated as follows:

(i) Choose a national life table for the standard population obtained in a year at about the middle of the study period.
(ii) With properly identified race, sex, entry age a_1, and duration t, we have approximately

$$r \cong (n_1 - n_2)/n_1$$

where n_1 and n_2 are the sizes (down from a hypothetical cohort of 100,000 at age zero) of the standard population at ages a_1 and $a_2 = a_1 + t$, respectively.

Example 7. Table 2.8 shows the summary of a cohort analysis. Two groups of male asbestos workers with different levels of exposure were followed up. The mortality was considered over a period starting 10 years after initial exposure, since the object was to evaluate the carcinogenic effect of asbestos exposure, and this effect does not result in death for at least 10 years.

TABLE 2.8. Mortality of Asbestos Workers

	Exposure Category			
	Low-to-Moderate		Severe	
Cause of Death	Observed	Expected	Observed	Expected
All causes	207	213.3	338	224.7
Cancer of lung and pleura	33	20.0	87	23.2
Gastrointestinal	19	16.3	39	17.7
Other cancers	14	13.2	32	14.2
Other causes	141	163.8	180	169.6

Data from Newhouse and Berry, 1979.

Model

Let's assume the proportional hazards model (PHM):

$$\lambda(t; \text{cohort}) = \rho \lambda(t; \text{standard population})$$

(Cox and Oakes, 1984). The constant ρ, called the *relative risk*, serves as a measure of excess mortality.

Results

Under this proportional hazards model, the individual expected mortalities r's may be treated as identically and independently distributed exponential variables with parameter ρ (the exponential hazard); a sketch of the proof is as follows:

(i) Let H be the cumulative hazard function:

$$H(t) = \int_0^t \lambda(x) dx$$

The model is expressible as

$$H(t; \text{cohort}) = \rho \cdot H(t; \text{standard population})$$

or

$$H(t) = \rho \cdot H_0(t)$$

(ii) When applied to the survival time T in the cohort, treated as a random variable, we have

$$H_0(T) = \frac{1}{\rho} H(T)$$

which is distributed as exponential with parameter ρ because

$$H(T) = -\ln S(T)$$

and $S(T)$ has the uniform distribution $U(0,1)$. Finally,

(iii) r is an estimate of $H_0(T)$ [w estimates $\lambda(t)$ and integration is replaced by a sum].

This leads to the following likelihood function in which d is the observed number of deaths (see Exponential Model, Section 2.2):

$$\ln L = -\rho e + d \ln \rho$$

(In this special application, the r's are "duration" times and e is their total.)

From the above likelihood function, we have:

(i) The constant relative risk ρ can be estimated by

$$\hat{\rho} = d/e \qquad \text{SE}(\hat{\rho}) = \sqrt{d}/e$$

This ratio, d/e, of observed over expected deaths, possibly expressed as a percentage, is the SMR, a common measure in occupational mortality studies.

(ii) A 95 percent confidence interval for ρ is given by

$$[(d + 1 - 1.96\sqrt{d})/e, \ (d + 2 + 1.96\sqrt{d})/e]$$

This formula for a 95 percent confidence interval follows a suggestion by Ury and Wiggins (1985) to add 1 to left and 2 to right endpoints in order to improve probability coverage due to skewness of the sampling distribution of $\hat{\rho}$. An alternative to the approach of Ury and Wiggins is to write

$$\rho = \exp(\beta)$$

and confidence intervals for ρ are formed through the use of confidence intervals for β (also, see Le, 1990).

Referring to Table 2.8, the effect of low-to-moderate exposure to cancer of lung and pleura may be assessed as follows:

(i) SMR = 1.65.
(ii) A 95 percent confidence interval for ρ is (1.14, 2.31).

Example 8. Refer to the sample of low-risk end-stage renal disease patients of Appendix B. The study was undertaken to serve two purposes:

(i) To study treatment adequacy. Inadequacy of treatment may not be a surprise result, because hemodialysis is not expected to cancel the impact of renal failure even when there are no other factors besides the metabolic deficiencies caused by the diseased kidney. However, the results would help to establish a quantitative standard so as to compare the effects of renal failure itself with other supplemental morbidity co-factors such as chronic obstructive pulmonary disease or diabetes.
(ii) To investigate certain interesting survival characteristics such as additional effects on patient survival of factors such as race, sex, age, and dialysis duration. The use of cohort analysis and SMR will allow us to derive results that would not be possible with other approaches such as regression analyses because of different expected mortalities between subgroups.

For the group of 513 end-stage renal disease patients on hemodialysis of Appendix B and using the U.S. population in 1980—

the middle point of the enrollment period—as the standard, we have

$$d = 81 \text{ deaths}$$

$$e = 17.95$$

leading to an SMR of 451 percent and a 95 percent confidence interval for the relative risk of (3.59, 5.61). The discrepancy between hemodialysis survival and the general population survival is due to the metabolic deficiencies caused by the diseased kidney and the inadequacy of hemodialysis as a treatment. The relative risk of the renal failure itself is high ($\hat{\rho} = 4.51$) as compared to other morbidity cofactors such as cerebrovascular accident, chronic obstructive pulmonary disease, or cancer, and is second only to diabetes. However, the results also show some optimism indicating an increasing trend in survival due to the improvements of hemodialysis techniques over the years.

Excess mortality was observed for all races, both sexes, in all age groups, and at all durations; more details are shown in Table 2.9. The SMRs for whites and nonwhites are very compatible, being 451 and 459 percent, respectively. The relatively high SMR for females (513 percent) as compared to males (397 percent) is somewhat attributable to the lower mortality rates experienced by women in the general population. Generally, mortality ratios representing excess risk were highest at the youngest ages and formed a decreasing trend with increasing age (987 percent for less-than-20, 750 percent for 20–39, 573 percent for 40–59, and 413 percent for 60-plus age groups when both sexes are combined). The only exception to this trend is for females in the 40–59 age group. This decreasing trend may be somewhat surprising because it is generally believed that advanced age is one of the most important prognostic factors in survival of hemodialysis patients (or patients with any disease). The fact is that age is still a very important prognostic factor with the elderly showing poor survival; however, the discrepancy between these diseased elderly and their "peers" in the general population is smaller as compared to a similar gap for a younger age group. Standardized mortality ratios were also highest in the first year of treatment and formed a steep decreasing trend with increasing duration in a very consistent fashion for both

TABLE 2.9. Survival Characteristics of Low-Risk End-Stage Renal Disease Patients on Hemodialysis

SMR (%) by Race and Sex			
	Race		
Sex	White	Nonwhite	Total
Male	413	235	397
Female	491	753	513
Total	451	459	453

SMR (%) by Age and Sex				
	Age Group (yr)			
Sex	< 20	20–39	40–59	60+
Male	1053	772	373	371
Female	—	689	784	456
Total	987	750	573	413

SMR (%) by Duration and Sex				
	Duration (yr)			
Sex	< 1	1–2	2–3	3+
Male	1052	588	319	384
Female	1720	977	584	293
Total	1131	724	458	289

Le et al., 1991.

sexes. The overall figures ranged from 1131 percent for the first year to 289 percent for those in the hemodialysis program 3 years or more. This slightly decreasing risk model was not detected by studies of smaller size and/or shorter follow-up times.

Healthy Worker Effects

It is noted that in some industrial cohort studies, the overall death rate found may be substantially less than that of the general population. This phenomenon, known as the "healthy worker" effect, is most likely a consequence of a selection factor whereby workers are necessarily in good health at their time of entry into the work force. There are two strategies to overcome this:

COHORT STUDIES

(i) Study only cause-specific mortality.
(ii) Study the attenuation of this effect with the passage of time.

Example 9. Data in Table 2.10 are stratified into four categories defined by years since first exposure in order to illustrate this feature of the healthy worker effect. The SMRs for circulatory disease almost form a linear trend increase from 21.5 to 76.9 percent. The interpretation of this finding is that the effect of the initial selectivity gradually wears off. Nevertheless, there is a continuing selection bias for those who remain employed; the final figure (76.9 percent) is still below 100 percent. Thus the healthy worker effect is continuously operative throughout the study, in the sense that persons who continue employment, and therefore continue to accumulate exposures, have lower relative mortality rates due to their presumably better health. Elimination of the first few years of follow-up may be a good strategy for coping with this problem; in Example 7, mortality was considered starting 10 years after initial exposure.

TABLE 2.10. Mortality of Vinyl Chloride Workers by Years Since Entering the Industry

Cause of Death		Years Since Entering Industry				Total
		1–4	5–9	10–14	15+	
All causes	Obs.	34	55	74	230	393
	Exp.	91.0	87.5	98.5	244.3	521.2
	SMR	37.4	62.9	75.1	94.2	75.4
All cancers	Obs.	9	15	23	68	115
	Exp.	20.3	21.3	24.5	60.8	126.8
	SMR	44.5	70.6	94	111.8	90.7
Circulatory disease	Obs.	7	25	38	110	180
	Exp.	32.5	35.6	44.9	121.3	234.2
	SMR	21.5	70.3	84.7	90.7	76.9
Respiratory disease	Obs.	2	4	4	32	42
	Exp.	9.6	10.3	12.8	34.4	67.1
	SMR	20.9	38.8	31.3	93.0	62.6

Data from Fox and Collier, 1976.

2.4. ESTIMATION OF THE CUMULATIVE HAZARD

Recall that the hazard function can be expressed as

$$\lambda(t) = f(t)/S(t)$$

Estimating $\lambda(t)$ is essentially equivalent to the difficult problem of estimating the density function $f(t)$. Tanner and Wong (1984) proposed a method together with its application to model diagnostics and exploratory analysis; however, numerical implementation remains complex. An easier problem is estimating the cumulative hazard function

$$H(t) = \int_0^t \lambda(x)\,dx$$

In the general setup of the previous section, we have

$$H(t) = \sum_{t_i \leq t} \lambda_i$$

where λ_i is the discrete hazard at t_i. Thus,

$$\hat{H}(t) = \sum_{t_i \leq t} \hat{\lambda}_i$$

$$= \sum_{t_i \leq t} (d_i/n_i)$$

and

$$\widehat{\mathrm{Var}}[\hat{H}(t)] = s^2(t)$$

$$= \sum_{t_i \leq t} \frac{\hat{\lambda}_i(1-\hat{\lambda}_i)}{n_i}$$

$$= \sum_{t_i \leq t} \frac{d_i(n_i - d_i)}{n_i^3}$$

Note that from the method of the previous section we can form confidence intervals for $S(t)$; for example, the 95 percent confidence interval is

$$\hat{S}(t) \pm 1.96\{\widehat{\text{Var}[\hat{S}(t)]}\}^{1/2}$$

However, at small or large values of t, confidence limits may be outside of the range $[0, 1]$. This can be avoided by first considering a 95 percent confidence interval for $H(t)$:

$$\hat{H}(t) \pm 1.96 s(t)$$

and exponentiating the endpoints because

$$S(t) = \exp[-H(t)]$$

Another alternative estimator for $H(t)$ is

$$\tilde{H}(t) = \sum_{t_i \leq t} \left\{ \sum_{m=0}^{d_i - 1} \frac{1}{n_i - m} \right\}$$

(see Fleming et al., 1980); these two estimates—$\hat{H}(t)$ and $\tilde{H}(t)$—are only different when d_i's are large and only near the right tail where the n_i's are small. Of course, we can also use $-\ln \hat{S}(t)$ as an estimate of $H(t)$ where $\hat{S}(t)$ is the product-limit estimate of the survival function $S(t)$.

Exercise 2.2 gives an application of $\hat{H}(t)$ (or $\tilde{H}(t)$) in formulating an alternative estimator of the survival function.

2.5. RANK CORRELATION FOR BIVARIATE CENSORED DATA

This section presents an extension of the definition of Kendall's tau coefficient of correlation for use with bivariate censored data; the concept was first presented in Section 1.6. This extension maintains the same idea of counting the number of (known) concordances minus the number of (known) discordances, then dividing by the number of pairs (Breslow, 1970; Oakes, 1982).

Let the pair of true survival times be denoted by (T_{1i}, T_{2i}), for $i = 1, 2, \ldots$; these would be observed in the absence of censoring. It is assumed that these pairs are independent and identically distributed with a common bivariate distribution whose correlation is under investigation. The sequence of potential censoring times (C_{1i}, C_{2i}) is also assumed to be identically distributed and independent of (T_{1i}, T_{2i}); however, C_{1i} need not be independent of C_{2i}. In fact, in many applications, each pair may have the same potential censoring time, that is, $C_{1i} = C_{2i}$. The duration times consist of:

$$t_{1i} = \min(T_{1i}, C_{1i})$$

$$t_{2i} = \min(T_{2i}, C_{2i})$$

together with indications of whether each t_{1i} or t_{2i} is censored or uncensored. Define $X_{1i} = t_{1i}$ if uncensored, $X_{1i} = \infty$ otherwise, with X_{2i} defined similarly. Thus, X_{1i} may be thought of as the largest possible value of T_{1i} *consistent with the observation made*, and the event $(X_{1i} < t_{1j})$, equivalent to $(T_{1i} < C_{1i}, C_{1j}, T_{1j})$, means that T_{1i} is observed less than T_{1j}. Now, for each pair of subjects (i, j) with $i < j$ the scoring function θ_{ij} is defined as follows: Let $L_{ij} = 1$ if $X_{1i} < t_{1j}$, $L_{ij} = -1$ if $X_{1j} < t_{1i}$ and $L_{ij} = 0$ otherwise. Define M_{ij} similarly for the X_2 and t_2 variables and set $\theta_{ij} = L_{ij} M_{ij}$. Then $\theta_{ij} = 1$ if i and j are observed to be concordant, that is, if the ordering of the T_{1i} and T_{1j} is the same as the ordering of T_{2i} and T_{2j}, $\theta_{ij} = -1$ if i and j are observed to be discordant. The average over all pairs (i, j) is the Kendall tau:

$$\tau = \left(\sum_{i<j} \theta_{ij} \right) / \binom{n}{2}$$

Example 10. This example concerns the days of survival of closely matched and poorly matched skin grafts on the same person. Parts of the data are presented in Table 2.11, with plus sign (+) indicating censored observations; the complete data set can be found in Example 4 of Chapter 3. To complete the calculation of Kendall's tau, scores for each pair of subjects are assigned as in

TABLE 2.11. Days of Survival of Skin Grafts on the Same Person

Pair	A	B	C	D
Close match	37	19	57+	93
Poor match	29	13	15	26

TABLE 2.12. Scores for Data of Table 2.11

Pair	L_{ij}	M_{ij}	θ_{ij}
AB	−1	−1	1
AC	1	−1	−1
AD	1	−1	−1
BC	1	1	1
BD	1	1	1
CD	0	1	0
Totals			1

Table 2.12. The procedure yields:

$$\tau = \tfrac{1}{6}$$
$$= .167$$

indicating a weak, positive correlation between skin grafts on the same person (of course, it is a very small subsample where a change in one observation may affect the result substantially).

The next example involves a much larger sample.

Example 11. An application of the above method to the data in Appendix C yields:

$$\tau = .56$$

indicating a moderate correlation between durations of ventilating tubes in the left and right ears of the same child.

EXERCISES

2.1. Verify that in the absence of censoring, the product-limit estimator reduces to the empirical survival distribution function (= 1 − empirical cumulative distribution function)

$$\hat{S}(t) = \frac{1}{n}\sum_{i=1}^{n} I(t_i, t)$$

where

$$I(x,t) = \begin{cases} 0 & \text{if } x \leq t \\ 1 & \text{if } x > t \end{cases}$$

Show also that Greenwood's formula reduces to the usual variance estimate in this case, that is,

$$\widehat{\text{Var}}[\hat{S}(t)] = \frac{\hat{S}(t)[1-\hat{S}(t)]}{n}$$

2.2. Let $H(t)$ be the cumulative hazard function defined by

$$H(t) = \int_0^t \lambda(x)\,dx$$

Nelson (1972) proposed as an estimator for $H(t)$

$$\hat{H}(t) = \sum_{\text{all } t_i < t} \frac{d_i}{n_i}$$

leading to another estimator for the survival function,

$$\tilde{S}(t) = \exp[-\hat{H}(t)]$$

Use the first-order Taylor expansion of the logarithm or exponential function to show that $\hat{S}$ (the product-limit estimator) and this new $\tilde{S}$ are *practically* equivalent.

EXERCISES

2.3. Given this small data set:

$$9, 13, 13^+, 18, 23, 28^+, 31, 34, 45^+, 48, 161^+$$

set up a table to calculate $\hat{S}(t_i)$ and make a graph for $\hat{S}(t)$.

2.4. Consider a distribution with the hazard function

$$\lambda(t) = \theta e^t \quad \text{for} \quad t > 0 \quad \text{and} \quad \theta > 0$$

and an uncensored sample $\{t_i\}_{i=1}^n$. Derive the likelihood function for θ and its MLE if a closed-form solution exists.

2.5. Refer to the placebo group of the leukemia data set in Example 1. Calculate the mean survival time $(\bar{t})$ and compare the result to the area under the estimated survival curve.

2.6. Refer to the 6–MP group of the leukemia data set in Example 1. We found, for example,

$$\hat{S}(7) = .8067$$

$$\text{SE}[\hat{S}(7)] = .0869$$

Assuming now that the survival times are exponentially distributed, re-calculate $\hat{S}(7)$ and its standard error under this new assumption.

2.7. Refer to the leukemia data set of Example 1. Compute and compare the 95 percent confidence intervals for $S_1(10)$ and $S_2(10)$, where $S_1(10)$ and $S_2(10)$ are survival rates at 10 weeks for the placebo and 6–MP groups, respectively.

2.8. Refer to the kidney transplantation data set in Appendix A and assume that the survival times are exponentially distributed. Compute and compare the 95 percent confidence intervals for the mean survival time of the ALG and non-ALG groups.

2.9. Consider the survival model characterized by this piecewise-constant hazard function:

$$\lambda(t) = \begin{cases} \lambda_1 & \text{if } 0 \leq t < \pi_1 \\ \lambda_2 & \text{if } \pi_1 \leq t < \pi_2 \\ \lambda_3 & \text{if } \pi_2 \leq t \end{cases}$$

where π_1 and π_2 are fixed and known constants. Derive the maximum-likelihood estimators (MLEs) for λ_1, λ_2, and λ_3 and their standard errors from a sample: $\{(t_i, \delta_i)\}_{i=1}^{n}$.

2.10. Given this particular life table,

$[x_{i-1}, x_i]$	d_i	m_i	n_i
0–1	47	19	126
1–2	5	17	60
2–3	2	15	38
3–4	2	9	21
4–5	4	6	10

complete the table with a column for $\hat{S}(x_{i-1})$ and make a graph for $\hat{S}(t)$.

2.11. Write a program and submit it with the data in Appendix A. This program should provide estimates based on the actuarial method and include the following features:
(a) Seven shorter intervals at the beginning: 0–1, 1–2, 2–3, 3–4, 4–6, 6–12, 12–20 months; and 2 longer intervals at the end: 120–160, 160–240 months; middle intervals are 20 months long.
(b) Two separate curves: one for the ALG group and one for the non-ALG group. Any gross observation on the effect of ALG as

1985 Cohort

Primary Diagnosis	No. of Patients	One-Year Survival Rate	Standard Error
Diabetes	7,661	.7217	.0051
Hypertension	6,927	.7703	.0050
Glomerulonephritis	4,274	.8758	.0050
Cystic kidney	1,102	.9096	.0085
Other urologic	2,009	.7815	.0091
Other cause	1,644	.7538	.0104
Unknown cause	1,947	.7799	.0093
Cause missing	1,522	.7170	.0113
Total	27,086		

EXERCISES

a protective factor for kidney graph survival? Short-term/long-term effect?

(c) Add program instructions to obtain the graphs of $\ln \hat{S}(t)$ and $\hat{S}(t)$. Any gross observation to compare the shape of the curves versus that of an exponential distribution? Does your observation from the graph agree with the monotonicity of the hazard functions as listed in the computer output?

2.12. Write a program similar to that in Exercise 2.11 (use the same intervals) and submit it with the data in Appendix B. Compare the survival pattern versus that of kidney transplants in 2.11 using both the survival and hazard functions; also investigate the effect of sex (two separate curves: one for men, one for women).

2.13. The following table provides the number of ESRD patients and Kaplan–Meier one-year survival rates and their standard errors for the 1985 cohort. Using the 1980 cohort as standard, calculate the cause-adjusted one-year survival rate for this cohort and compare it to the cause-adjusted rate for the 1990 cohort, namely a z test.

2.14. The following table presents the results of a mortality study of male workers employed at an East London asbestos factory. This cohort began work between April 1, 1933, and March 31, 1964; the factory closed in 1968.

	Exposure category							
	Low to Moderate				Severe			
	< 2 Years (884)		> 2 Years (554)		< 2 Years (937)		> 2 Years (512)	
Cause of Death	Obs.	Exp.	Obs.	Exp.	Obs.	Exp.	Obs.	Exp.
All causes	118	118.0	89	95.3	162	122.2	176	102.5
Cancers of lung and pleura (ICD 162, 163)	17	11.1	16	9.0	31	12.8	56	10.4
Gastrointestinal cancer (ICD 150–158)	10	9.0	9	7.3	20	9.5	19	8.2
Other cancers	6	7.4	8	5.8	16	7.9	16	6.3
Chronic respiratory disease	19	17.5	16	14.7	20	17.6	28	15.9

With an emphasis on chronic respiratory disease for subjects with more than 2 years of exposure,

(a) Investigate the excess risk for severe exposure.

(b) Compare the levels of excess risk for the two groups: low-to-moderate versus severe exposure (through the use of confidence intervals; e.g., to see if the confidence intervals are overlapsed).

2.15. Refer to the survival times of the subjects with low-fat and saturated-fat diets in Example 13 of Chapter 1. Constructing necessary graph or graphs that can be used to check whether the two groups have proportional hazards; and if they do, give a rough estimate of the relative risk of saturated-fat versus low-fat diets.

2.16. Refer to the hemodialysis data set in Appendix B:

(a) Graph the two survival curves, one for men and one for women, of Exercise 2.12 on the log scale. Any gross observation on whether survival times fitting the exponential distributions?

(b) Assume now that the survival times are exponentially distributed, compute and compare the 95 percent confidence intervals for the mean survival time for men and women, for white and nonwhites.

2.17. Consider a sample exponential regression model with one covariate Z. Let $x_i = z_i - \bar{x}$, where z_i is the covariate value for the ith subject, and assume the model

$$\lambda_i = \frac{1}{a} \exp(-bx_i)$$

Suppose there is no censoring; verify that $\hat{a}$ and $\hat{b}$ are uncorrelated and find Var($\hat{a}$) in this case (Feigl and Zelen, 1965, Glasser, 1967).

2.18. Consider the two-sample problem with an exponential regression model as in equation (A) of Exercise 2.17 with

$$x = \begin{cases} 0 & \text{for group 1} \\ 1 & \text{for group 2} \end{cases}$$

Prove that if d is the number of events and V is the total duration time, then we have

$$\hat{b} = -\ln \frac{V_1 d_2}{V_2 d_1} \qquad \text{Var}(\hat{b}) = (d_1 + d_2)/d_1 d_2$$

EXERCISES

2.19. Refer to the cure models of Exercise 1.6 and the design to compare two smoking cessation programs. Indicate how to estimate the difference of treatment effects:

$$\theta = p_{AM} - p_{NM}$$

2.20. Refer to the organ transplant model of Exercise 1.9 where the survival time T of an individual depends on an unobservable transition time X during which the new organ may be rejected; X itself is a random variable with density $f(x)$. We may consider

$$\theta = \Pr(T \geq X)$$

as a "transplant success index." Assume that $f(x)$ is a gamma density.
(a) Indicate how to obtain the MLE for θ.
(b) Let $z = 0/1$ be an indicator for an immune-suppression drug ($z = 1$ if drug is used, $z = 0$ otherwise). Outline a strategy to investigate the drug effect through the use of a regression model such as

$$\theta = e^{\beta z}$$

CHAPTER 3

Comparison of Survival Distributions

3.1. Nonparametric Methods
   Comparison of Two Groups
   Comparison of Several Groups
   Detection of Crossing-Curves Alternatives
   Analysis of Pair-Matched Data
3.2. Comparison of Two Exponential Distributions
   Score Test
   Cox F Test
   Sample Size Determination
   Methods for Person-Years Data
3.3. Piecewise Exponential: Link Between Parametric and Nonparametric Methods
3.4. Trends in Survival
   Class of Tests Against Stochastically Ordered Alternatives
   Tarone Test
   Le–Gramsch–Louis Method
   Test for Trend in Constant Hazards and Its Application in Occupational Health
3.5. Comparison of Two Clustered Samples
   Problem
   Data Summarization
   Models
   Test Statistic
Exercises

The problem of comparing survival distributions arises very often in biomedical research. A laboratory researcher may want to compare the tumor-free times of two or more groups of rats exposed to carcinogens. A medical researcher may wish to compare the retinopathy-free times of groups of diabetic patients. A clinical oncologist may be interested in comparing the ability of two or more treatments to prolong life or maintain health. An epidemiologist may be interested in evaluating a public health intervention or an occupational hazard or may wish to compare the exposed versus nonexposed groups with respect to deaths due to cardiovascular diseases or cancers. The Multiple Risk Factor Intervention Trial (MRFIT), for example, provides an example of such a public health intervention. The MRFIT is one of the largest randomized clinical trials ever conducted, in which over 360,000 middle-aged men were screened at 22 locations around the United States to identify 12,866 men who were the participants in the trial itself. The MRFIT had as its primary aim to ascertain whether an intervention program directed to achieve cigarette smoking cessation, lowering of blood pressure, and lowering blood cholesterol levels would result in a reduction in deaths from coronary heart disease (Multiple Risk Factor Intervention Trial Research Group, 1982).

In this as well as other examples, a rough idea of the differences between groups can be illustrated by drawing graphs of the estimated survival curves; however, real differences can only be revealed by application of statistical tests of significance. Section 3.1 deals with nonparametric methods and Section 3.2 is concerned with comparison of two exponential samples with applications to sample size determination and to the analysis of epidemiological cohort studies. Section 3.3 shows the relationship between parametric and nonparametric methods. Section 3.4 presents tests against trend in survival. Section 3.5 presents a special health application with clustered survival data.

3.1. NONPARAMETRIC METHODS

When there are no censored observations, standard nonparametric tests can be used to compare survival distributions; for example, the Wilcoxon or Mann–Whitney for the comparison of two samples, and the Kruskal–Wallis test for the comparison of several groups (see Chapter 1). In this section we introduce a family of nonparametric tests for samples with censoring, called the *Tarone–Ware* class of tests

(Tarone and Ware, 1977; Gehan, 1965a; Breslow, 1970; Cox, 1972), a member of which reduces to the Wilcoxon or Kruskal–Wallis test in the absence of censoring.

Comparison of Two Groups

Suppose that there are n_1 and n_2 individuals corresponding to two treatment groups 1 and 2, respectively. The study provides two samples of survival data:

$$\{(t_{1i}, \delta_{1i})\} \qquad i = 1, 2, \ldots n_1$$

and

$$\{(t_{2j}, \delta_{2j})\} \qquad j = 1, 2, \ldots n_2$$

In the presence of censored observations, tests of significance can be constructed as follows:

(i) Pool data from two samples together and let

$$t_1 < t_2 < \cdots < t_m \qquad m \leq d \leq n_1 + n_2$$

be the distinct times with at least one event at each (d is the total number of deaths).

(ii) At ordered time t_i, $1 \leq i \leq m$, the data may be summarized into a 2×2 table:

Sample	Dead	Alive	Total
1	d_{1i}	a_{1i}	n_{1i}
2	d_{2i}	a_{2i}	n_{2i}
Total	d_i	a_i	n_i

(Status: Dead, Alive)

where n_{1i} = number of subjects from sample 1 who were at risk just before time t_i

n_{2i} = number of subjects from sample 2 who were at risk just before time t_i

$n_i = n_{1i} + n_{2i}$

d_i = number of deaths at t_i, d_{1i} of them from sample 1 and d_{2i} of them from sample 2

$\quad = d_{1i} + d_{2i}$

$a_i = n_i - d_i$

$\quad = a_{1i} + a_{2i}$

$\quad$ = number of survivors

$d = \sum d_i$

In this form, the null hypothesis of equal survival functions implies the independence of "sample" and "status" in the above cross-classified 2×2 table. Therefore, under the null hypothesis, the "expected value" of d_{1i} is

$$E_0(d_{1i}) = \frac{n_{1i} d_i}{n_i}$$

(d_{1i} being the "observed value"). The variance is estimated by (hypergeometric model):

$$\text{Var}_0(d_{1i}) = \frac{n_{1i} n_{2i} a_i d_i}{n_i^2 (n_i - 1)}$$

After constructing a 2×2 table for each uncensored time, the evidence against the null hypothesis can be summarized in the following statistic:

$$\theta = \sum_{i=1}^{m} w_i [d_{1i} - E_0(d_{1i})]$$

where w_i is the "weight" associated with the 2×2 table at t_i. We have under the null hypothesis:

$$E_0(\theta) = 0$$

$$\text{Var}_0(\theta) = \sum_{i=1}^{m} w_i^2 \, \text{Var}_0(d_{1i})$$

$$= \sum_{i=1}^{m} \frac{w_i^2 n_{1i} n_{2i} a_i d_i}{n_i^2 (n_i - 1)}$$

The evidence against the null hypothesis is summarized in the standardized statistic

$$\boxed{z = \theta / \{\text{Var}_0(\theta)\}^{1/2}}$$

which is referred to the standard normal percentile $z_{1-\alpha}$ for a specified size α of the test. We may also refer z^2 to a chi-square distribution at 1 degree of freedom.

There are two important special cases:

(i) The choice

$$w_i = n_i$$

gives the generalized Wilcoxon test (also called the Gehan–Breslow test), (Gehan, 1965a and 1965b; Breslow, 1970); it is reduced to the Wilcoxon test in the absence of censoring.

(ii) The choice

$$w_i = 1$$

gives the log-rank test (also called the Cox–Mantel test; it is similar to the Mantel–Haenszel procedure for the combination of several 2 × 2 tables in the analysis of categorical data) (Mantel and Haenszel, 1959; Cox, 1972; Tarone and Ware, 1977).

Other choices are possible, for example $w_i = \sqrt{n_i}$ (Tarone–Ware test). Another possibility is Peto's, which uses as the weight the estimate of the survival function from pooled data; this is similar to the generalized Wilcoxon, but the weights are not affected by censoring.

The log-rank test (with weight $w_i = 1$) can also be derived as follows. Let

$$X = \begin{cases} 1 & \text{for a subject in sample 1} \\ 0 & \text{for a subject in sample 2} \end{cases}$$

be the group indicator and assume the proportional hazards model

$$\lambda(t) = \lim_{\delta \downarrow 0} \frac{\Pr[t \leq T \leq t + \delta \mid t \leq T]}{\delta}$$

$$= \lambda_0(t) e^{\beta x}$$

as introduced in Section 1.4. Denote the ordered distinct death times by

$$t_1 < t_2 < \cdots < t_m$$

and let R_i be the risk set just before time t_i, n_i the number of subjects in R_i, D_i the death set at time t_i, d_i the number of subjects (i.e., deaths) in D_i, and C_i the collection of all possible combinations of subjects from R_i; each combination—or subset of R_i—has d_i members, and D_i is itself one of these combinations. For example, if three subjects (A, B, and C) are at risk just before time t_i and two of them (A and B) die at t_i, then

$$R_i = \{A, B, C\} \qquad n_i = 3$$
$$D_i = \{A, B\} \qquad d_i = 2$$
$$C_i = \{\{A, B\} = D_i, \{A, C\}, \{B, C\}\}$$

NONPARAMETRIC METHODS

The number of elements in C_i is

$$\binom{n_i}{d_i}$$

Cox (1972) suggests using

$$L = \prod_{i=1}^{m} \Pr(D_i \mid R_i, d_i)$$
$$= \prod_{i=1}^{m} \frac{\exp(\beta s_i)}{\sum_{C_i} \exp(\beta s_u)}$$

as a likelihood function in which

$$s_i = \sum_{D_i} x_j$$

$$s_u = \sum_{D_u} x_j \qquad D_u \in C_i$$

The score test based on this proportional hazards model

$$z = \left\{ \frac{d}{d\beta} \log L \text{ at } \beta = 0 \right\} \Big/ \left\{ -\frac{d^2}{d\beta^2} \log L \text{ at } \beta = 0 \right\}^{1/2}.$$

is the above log-rank procedure; details are in Chapter 4. The generalized Wilcoxon test (with weight $w_i = n_i$) can be derived in two different ways:

(i) Gehan (1965a) introduced the test as a direct generalization of the U statistic. Define

$$U_{ij} = \begin{cases} +1 & \text{if } t_{1i} > t_{2j} \text{ and } \delta_{2j} = 1 \\ -1 & \text{if } t_{1i} < t_{2j} \text{ and } \delta_{1i} = 1 \\ 0 & \text{otherwise} \end{cases}$$

Then it can be shown that the generalized Wilcoxon (or Gehan's) test is based on

$$\sum_{i=1}^{n_1}\sum_{j=1}^{n_2} U_{ij}$$

Of course, it is reduced to the Wilcoxon rank-sum test in the absence of censoring.

(ii) Another alternative approach is to pool the data to form a sample of size $(n_1 + n_2)$; then arrange all the times (including censored ones) from smallest to largest and let n_j be the number of subjects at risk at time t_j and

$$s_i = \prod_{j=1}^{i} \frac{n_j}{n_j + 1}$$

A "score" is then assigned to each subject as follows:

(i) If the subject is the ith event, the score is

$$c_i = 1 - 2s_i$$

(ii) If the subject is censored between the ith and the $(i + 1)$th events, the score is

$$c_i = 1 - s_i$$

The generalized Wilcoxon test can be formed by adding scores from group 1 (Mantel, 1967).

There are a few other interesting issues:

1. Which test should we use? The generalized Wilcoxon statistic puts more weight on the beginning observations and because of that its use is more powerful in detecting the effects of short-term risks. On the other hand, the log-rank statistic puts equal weight on each observation and therefore, by default, is more sensitive to exposures with a constant relative risk (proportional hazards effect; in fact, we have derived the log-rank test as a

score test using the proportional hazards model). Because of these characteristics, applications of both tests may reveal not only whether or not an exposure has any effect but also the nature of the effect, short-term or long-term.

2. Because of the way the tests are formulated (terms in the summation are not squared),

$$\sum_{\text{all } i} w_i [d_{1i} - E_0(d_{1i})]$$

they are only powerful when one risk is greater than the other at all times. Otherwise, some terms in this sum are positive, some other terms are negative, and they cancel each other out. For example, the tests are virtually powerless for the case of crossing hazards (see Example 2); in this case the assumption of proportional hazards is severely violated.

3. Some cancer treatments (bone marrow transplantation, e.g.) are thought to have cured patients within a short time of initiation. Then, instead of all patients having the same hazard, a biologically more appropriate model, the cure model, assumes that an unknown proportion $(1 - \pi)$ are still at risk whereas the remaining proportion (π) have essentially no risk. If the aim of the study is to compare the cure proportions π's, then neither the generalized Wilcoxon nor log-rank tests are appropriate (low power). One may simply choose a time point t far enough for the curves to level off, then compare the estimated survival rates by referring to percentiles of the standard normal distribution:

$$z = \frac{\hat{S}_1(t) - \hat{S}_2(t)}{\{\text{Var}[\hat{S}_1(t)] + \text{Var}[\hat{S}_2(t)]\}^{1/2}}$$

Estimated survival rates, $\hat{S}_i(t)$, and their variances are obtained as in Chapter 2 (Kaplan–Meier procedure).

Example 1. Refer back to the clinical trial to evaluate the effect of 6–mercaptopurine (6–MP) to maintain remission from acute leukemia (Example 1 of Chapter 2). The results of the tests indicate a highly significant difference between survival patterns of the two groups. The generalized Wilcoxon test shows a slightly larger statistic indicating that the difference is slightly larger at earlier times; however, the log-rank test is almost equally significant indicating that the use of 6–MP has a long-term effect (the effect does not wear off).

$$\text{Generalized Wilcoxon:} \quad \chi^2 = 13.46 \quad (1\ df); \quad p < .0001$$
$$\text{Log-rank:} \quad \chi^2 = 16.79 \quad (1\ df); \quad p = .0002$$

Note: An SAS program would include these instructions: PROC LIFETEST METHOD = KM; TIME WEEKS*RELAPSE(0); STRATA DRUG; where KM stands for Kaplan–Meier method, WEEKS is the variable name for duration time, RELAPSE the variable name for survival status, "0" is the coding for censoring, and DRUG is the variable name specifying groups to be compared (this grouping variable may have more than two levels).

The next example shows a type of departure from the null hypothesis for which both tests, the log-rank and the generalized Wilcoxon, may fail to detect.

Example 2. In a study performed at the Mayo Clinic (data from Fleming et al., 1980), patients with bile duct cancer were followed to determine whether those treated with a combination of radiation treatment (RöRx) and 5-Fluorouracil (5-Fu) would survive significantly longer than a control population.

Survival times for the RöRx + 5-Fu sample were 30, 67, 79$^+$, 82$^+$, 95, 148, 170, 171, 176, 193, 200, 221, 243, 261, 262, 263, 399, 414, 446, 446$^+$, 464, and 777 days.

Survival times for the control sample were 57, 58, 74, 79, 89, 98, 101, 104, 110, 118, 125, 132, 154, 159, 188, 203, 257, 257, 431, 461, 497, 723, 747, 1313, and 2636 days.

Estimates of the survival curves are presented in Figure 3.1. It is obvious that both the log-rank and the generalized Wilcoxon tests are insensitive to this type of departure from the null hypothesis, yielding p values of .418 and .127, respectively (these authors developed a "generalized Smirnov procedure," which yields a significance $p = .046$. This testing procedure is rather complicated; later in this section, we will introduce a more simple approach for the detection of "crossing-curves alternatives."

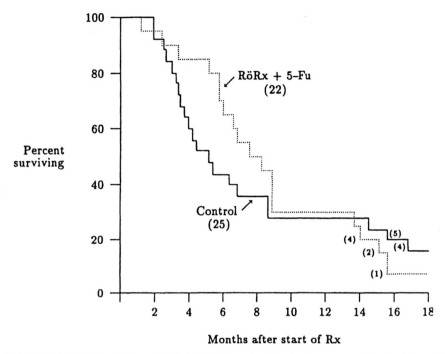

FIGURE 3.1. Estimated survival distributions for bile duct cancer patients: Control population vs. patients receiving RöRx+5-Fu.

Comparison of Several Groups

Methods of the previous subsection on comparison of two groups can be generalized to simultaneously compare k treatment groups ($k > 2$). Ordered times of events are formed:

$$t_1 < t_2 < \cdots < t_m$$

and at t_i, data are summarized into a $2 \times k$ table:

Status	\multicolumn{5}{c}{Sample}	Total
	1	2	$\cdots$	j	$\cdots$	k	Total
Dead	d_{1i}	d_{2i}	$\cdots$	d_{ji}	$\cdots$	d_{ki}	d_i
At risk	n_{1i}	n_{2i}	$\cdots$	n_{ji}	$\cdots$	n_{ki}	n_i

Under the null hypothesis, the expected value of d_{ji} is

$$E_0(d_{ji}) = \frac{n_{ji} d_i}{n_i}$$

The evidence against $\mathcal{H}_0$ can be summarized into

$$\boxed{\theta = \sum_{i=1}^{m} w_i S_i}$$

where w_i is the weight associated with the observations at t_i and S_i is the following random vector measuring the difference between observed and expected numbers of deaths:

$$S_i = \begin{bmatrix} d_{1i} - E_0(d_{1i}) \\ \vdots \\ d_{ji} - E_0(d_{ji}) \\ \vdots \\ d_{ki} - E_0(d_{ki}) \end{bmatrix}$$

The use of the statistic $\sum w_i S_i$ leads to a chi-square test with $(k-1)$ degrees of freedom. There are still two popular choices (among others) for the weights:

(i) The choice

$$w_i = n_i$$

gives the generalized Wilcoxon test (or generalized Kruskal–Wallis; also called the Gehan–Breslow test). Another alternative is Peto's weight.

(ii) The choice

$$w_i = 1$$

gives the log-rank test (or Cox–Mantel test).

NONPARAMETRIC METHODS 109

The relationship between these tests remains the same: The Gehan–Breslow test is more powerful in detecting the effects of short-term risks whereas the Cox–Mantel test is more sensitive to differences later in time. Both are weak for the cases of crossing hazards.

Similar to the case of two samples, the conditional distribution of $\{d_{1i}, d_{2i}, \ldots, d_{ki}\}$ under $\mathcal{H}_0$ is (multivariate) hypergeometric. The mean of d_{ji} under the null hypothesis is

$$E_0(d_{ji}) = d_i n_{ji}/n_i$$

and components of the covariance matrix V are

$$\text{Var}_0(d_{ji}) = n_{ji}(n_i - n_{ji})d_i a_i / n_i^2(n_i - 1)$$

$$\text{Cov}_0(d_{ji}, d_{li}) = -n_{ji} n_{li} d_i a_i / n_i^2(n_i - 1)$$

The test statistic

$$\chi^2 = \theta' V_w^{-1} \theta \qquad V_w = w^2 V$$

is distributed as chi-square with $(k-1)$ degrees of freedom under $\mathcal{H}_0$.

Detection of Crossing-Curves Alternatives

Let us consider again the data of Example 2:

Treated sample: 30, 67, 79$^+$, 82$^+$, 95, 148, 170, 171, 176, 193, 200, 221, 243, 261, 262, 263, 399, 414, 446, 446$^+$, 464, 777.
Control sample: 57, 58, 74, 79, 89, 98, 101, 104, 110, 118, 125, 132, 154, 159, 188, 203, 257, 257, 431, 461, 497, 723, 747, 1313, and 2636.

It can be seen from Figure 3.1 that the survival curves crossed and both the log-rank and the generalized Wilcoxon tests are insensitive to this type of crossing-curves alternative. A simple approach to this problem is as follows:

(i) Pool the data to form a sample of size $(n_1 + n_2)$; then arrange all the uncensored times from smallest to largest; let n_j be the number of subjects at risk at time t_j and

$$s_i = \prod_{j=1}^{i} \frac{n_j}{n_j + 1}$$

(ii) A score c_i is assigned to each observation:
   (a) If the subject is the ith event,

$$c_i = 1 - 2s_i$$

   (b) If the subject is censored between the ith and $(i+1)$th events,

$$c_i = 1 - s_i$$

(iii) Let $\bar{c}_1$ and $\bar{c}_2$ be the average score for groups 1 and 2, respectively. For each subject in each of the two groups, compute the absolute value of the deviation from average score

$$d = c - \bar{c}$$

(iv) Apply the Wilcoxon rank sum test to these d_i's; the idea is similar to that of Levene (Levene, 1960).

Example 3. To apply the above procedure to the data of Example 2, we have the following table.

We have the sum of the ranks for the treated sample:

$$R = 430$$

NONPARAMETRIC METHODS

Under $\mathcal{H}_0$, this sum has mean and standard deviation

$$\mu = \tfrac{1}{2}(22)(22+25+1)$$
$$= 528$$

Treated Sample				Control Sample			
Time	Score	|Deviation|	Rank	Time	Score	|Deviation|	Rank
30	−.958	1.047	47	57	−.916	.820	41
67	−.834	.923	44	58	−.874	.778	39
79+	.125	.036	1.5	74	−.792	.696	36
82+	.125	.036	1.5	79	−.750	.654	33
95	−.662	.751	37	89	−.706	.610	32
148	−.312	.401	25	98	−.678	.582	30
170	−.182	.271	17	101	−.574	.478	28
171	−.138	.227	14	104	−.532	.436	26
176	−.094	.183	12	110	−.488	.392	24
193	−.006	.095	7	118	−.444	.348	22
200	.038	.051	5	125	−.400	.304	18
221	.126	.037	3	132	−.356	.260	15
243	.170	.081	6	154	−.268	.172	10
261	.260	.171	9	159	−.224	.128	8
262	.306	.217	13	188	−.050	.046	4
263	.352	.263	16	203	.082	.178	11
399	.398	.309	21	257	.212	.308	19.5
414	.444	.355	23	257	.212	.308	19.5
446	.538	.449	27	431	.490	.586	31
446+	.769	.680	34	461	.588	.684	35
464	.640	.551	29	497	.692	.788	40
777	.846	.757	38	723	.742	.838	42
				747	.794	.890	43
				1313	.898	.994	45
				2636	.948	1.044	46

and

$$\sigma = \{\tfrac{1}{12}(22)(25)(22+25+1)\}^{1/2}$$
$$= 46.9$$

respectively. These lead to a z score of

$$z = \frac{430-528}{46.9}$$
$$= -2.09 \qquad (p = .018)$$

which is similar to the result of Fleming et al. (1980) (see Example 2).

Analysis of Pair-Matched Data

Twin studies and pair-matched experiments permit the study of the relation between a response and factors of interest in a manner that will often be relatively free from the effects of extraneous factors. For the case of survival data, commonly used nonparametric procedures are often inadequate; the sign test is insensitive because it ignores the magnitude of the difference and the Wilcoxon's signed rank test has been shown to be too sensitive to outliers, especially when the subject (or the pair, rather than the individual measurement) is the cause of the outlier (the procedure is also complicated in the presence of censoring).

Suppose we have n pairs of survival times with common (right) potential censoring times for both pair members. A procedure known as the paired Prentice–Wilcoxon test has been proposed (Holt and Prentice, 1974), consisting of the following steps.

(i) Ignore the matching and pool the data to form a right-censored sample of size $2n$.
(ii) Arrange all the observation times from smallest to largest and let n_j be the number of grafts at risk at time t_j and define

$$s_i = \prod_{j=1}^{i} \frac{n_j}{n_j + 1}$$

(iii) A score is assigned to each subject as follows:
- If the subject is the ith observed death, the score is

$$c_i = 1 - 2s_i$$

- If the subject is censored between the ith and $(i+1)$th deaths, the score is

$$c_i = 1 - s_i$$

(iv) Let the scores for the lth pair be (c_{1l}, c_{2l}); then define

$$\Delta_l = c_{1l} - c_{2l}$$

and

$$z = \sum_{l=1}^{n} \Delta_l \bigg/ \left\{ \sum_{l=1}^{n} \Delta_l^2 \right\}^{1/2}$$

Under the null hypothesis of no difference between the two members of pairs, z has a standard normal distribution (this is referred to as the "paired t test on the ranks" procedure). The scoring here is very much the same procedure as in the generalized Wilcoxon test and the basic idea is to apply a paired t test to these scores (Conover and Iman, 1981; O'Brien and Fleming, 1987).

Example 4. This example concerns the days of survival of closely matched and poorly matched skin grafts on the same person. Data are presented in Table 3.1, with plus sign (+) indicating censored observations.

TABLE 3.1. Days of Survival of Skin Grafts on the Same Person

Pair	1	2	3	4	5	6	7	8	9	10	11
Close match	37	19	57+	93	16	22	20	18	63	29	60+
Poor match	29	13	15	26	11	17	26	4	43	15	40

TABLE 3.2. Scores for Data of Table 3.1

	Times		Scores			
Pair	Close Match (1)	Poor Match (2)	c_{1i}	c_{2i}	Δ	Δ^2
1	37	29	.208	.094	.114	.013
2	19	13	−.372	−.826	.454	.206
3	57+	15	.717	−.740	1.457	2.123
4	93	26	.812	−.006	.818	.669
5	16	11	−.646	−.912	.266	.071
6	22	17	−.098	−.556	.458	.210
7	20	26	−.740	−.006	−.734	.539
8	18	21	−.464	−.190	−.274	.075
9	63	43	.622	.434	.188	.035
10	29	15	.094	−.740	.834	.696
11	60+	40	.717	.320	.397	.158
Totals					3.978	4.795

To complete the Prentice–Wilcoxon procedure, pool the data; scores are assigned as in Table 3.2. The procedure yields $z = 1.827$ ($p = .017$; no packaged program is available yet)

$$z = \sum \Delta \Big/ \left\{ \sum \Delta^2 \right\}^{1/2}$$

$$= 3.978/\sqrt{4.795}$$

$$= 1.827$$

3.2. COMPARISON OF TWO EXPONENTIAL DISTRIBUTIONS

Suppose again that there are n_1 and n_2 individuals corresponding to two treatment groups 1 and 2, respectively. The study provides two independent samples of survival data:

$$\{(t_{1i}, \delta_{1i})\}_{i=1}^{n_1} \quad \text{and} \quad \{t_{2i}, \delta_{2i})\}_{i=1}^{n_2}$$

If it is known that the survival times of the two groups follow the exponential distribution with hazards λ_1 and λ_2, respectively, then testing the equality of two exponential distributions is equivalent to testing

$$\mathcal{H}_0 : \lambda_1 = \lambda_2$$

Score Test

Let

$$T_i = \sum_{j=1}^{n_i} t_{ij}$$

and

$$d_i = \sum_{j=1}^{n_i} \delta_{ij}$$

be the total duration and number of events in group i; $i = 1,2$. The joint likelihood is given by

$$\ln L = -\lambda_1 T_1 + d_1 \ln \lambda_1 - \lambda_2 T_2 + d_2 \ln \lambda_2$$

Let

$$\lambda_1 = \exp(\beta_1)$$
$$\lambda_2 = \exp(\beta_1 + \beta_2)$$

so that $\mathcal{H}_0$ is expressible as $\mathcal{H}_0 : \beta_2 = 0$ and

$$\ln L = -T_1 \exp(\beta_1) - T_2 \exp(\beta_1 + \beta_2) + \beta_1 d_1 + (\beta_1 + \beta_2) d_2$$

Under $\mathcal{H}_0$, β_1 is estimated by

$$\exp(\hat{\beta}_1) = (d_1 + d_2)/(T_1 + T_2)$$

Therefore

$$\frac{\delta \ln L}{\delta \beta_1} = -T_1 \exp(\beta_1) - T_2 \exp(\beta_1 + \beta_2) + d_1 + d_2$$

$$\stackrel{\hat{}}{\underset{\mathcal{H}_0}{=}} 0$$

$$\frac{\delta \ln L}{\delta \beta_2} = -T_2 \exp(\beta_1 + \beta_2) + d_2$$

$$\stackrel{\wedge}{\underset{\mathcal{H}_0}{=}} (d_2 T_1 - d_1 T_2)/(T_1 + T_2)$$

$$-\frac{\delta^2 \ln L}{\delta \beta_1^2} = T_1 \exp(\beta_1) + T_2 \exp(\beta_1 + \beta_2)$$

$$\stackrel{\wedge}{\underset{\mathcal{H}_0}{=}} (d_1 + d_2)$$

$$-\frac{\delta^2 \ln L}{\delta \beta_1 \delta \beta_2} = T_2 \exp(\beta_1 + \beta_2)$$

$$\stackrel{\wedge}{\underset{\mathcal{H}_0}{=}} \frac{T_2(d_1 + d_2)}{T_1 + T_2}$$

$$-\frac{\delta^2 \ln L}{\delta \beta_2^2} = T_2 \exp(\beta_1 + \beta_2)$$

$$\stackrel{\wedge}{\underset{\mathcal{H}_0}{=}} \frac{T_2(d_1 + d_2)}{T_1 + T_2}$$

From these results we have

$$\chi^2_{ES} = \begin{bmatrix} 0 & \dfrac{d_2 T_1 - d_1 T_2}{T_1 + T_2} \end{bmatrix} \begin{bmatrix} d_1 + d_2 & \dfrac{T_2(d_1 + d_2)}{T_1 + T_2} \\ \dfrac{T_2(d_1 + d_2)}{T_1 + T_2} & \dfrac{T_2(d_1 + d_2)}{T_1 + T_2} \end{bmatrix}^{-1} \begin{bmatrix} 0 \\ \dfrac{d_2 T_1 - d_1 T_2}{T_1 - T_2} \end{bmatrix}$$

$$= \begin{bmatrix} 0 & \dfrac{d_2 T_1 - d_1 T_2}{T_1 + T_2} \end{bmatrix} \begin{bmatrix} \dfrac{T_1 + T_2}{d_1(d_1 + d_2)} & -\dfrac{T_1 + T_2}{d_1(d_1 + d_2)} \\ -\dfrac{T_1 + T_2}{d_1(d_1 + d_2)} & \dfrac{(T_1 + T_2)^2}{T_1 T_2(d_1 + d_2)} \end{bmatrix} \begin{bmatrix} 0 \\ \dfrac{d_2 T_1 - d_1 T_2}{T_1 + T_2} \end{bmatrix}$$

$$\boxed{\chi^2_{ES} = \frac{(d_2 T_1 - d_1 T_2)^2}{T_1 T_2 (d_1 + d_2)}}$$

Taking the square root we have

$$z = \frac{d_2 T_1 - d_1 T_2}{\sqrt{T_1 T_2 (d_1 + d_2)}}$$

$$= \frac{d_2 - (d_1 + d_2)\left(\frac{T_2}{T_1 + T_2}\right)}{\sqrt{(d_1 + d_2)\left(\frac{T_2}{T_1 + T_2}\right)\left(1 - \frac{T_2}{T_1 + T_2}\right)}}$$

$$= \frac{d_2 - n\pi}{\sqrt{n\pi(1 - \pi)}}$$

corresponding to a normal approximation to the conditional binomial distribution $B(n, \pi)$ of d_2 under $\mathcal{H}_0$ where

$$n = d_1 + d_2$$

$$\pi = T_2 / (T_1 + T_2)$$

(i.e., numbers of deaths are proportional to the total exposure times under $\mathcal{H}_0$).

Example 5. For the data set on leukemia patients of Example 1 from Chapter 2, the numbers of deaths and total exposure times are:
6–MP group:

$$d_1 = 9$$

$$T_1 = 359$$

Placebo group:

$$d_2 = 21$$

$$T_2 = 182$$

Therefore, if exponential models are assumed, the comparison of the two groups is based on the efficient score chi-square

$$\chi^2_{ES} = \frac{[(21)(359) - (9)(182)]^2}{(359)(182)(9+21)}$$

$$= 17.76$$

showing a significant drug effect.

Cox F Test

Within the framework of the above conditional binomial distribution, the level of significance (or p value) for testing $\mathcal{H}_0$ against $\mathcal{H}_A$ is given by

$$p = \frac{1}{2} \left\{ \sum_{u=0}^{d_1} \binom{n}{u} \pi^u (1-\pi)^{n-u} + \sum_{v=0}^{d_1-1} \binom{n}{v} \pi^v (1-\pi)^{n-v} \right\} \quad (1)$$

Now we have:

(i) If p_1 is such that

$$p_1 = \sum_{u=0}^{d_1} \binom{n}{u} \pi^u (1-\pi)^{n-u}$$

then

$$T_1/T_2 = (v'_1 F_{v'_1, v'_2, 1-p_1})/v'_2$$

(see Johnson and Kotz, 1969, pp. 58–59) where n, π, T_1, and T_2 are as previously defined,

$$v'_1 = 2(d_1 + 1)$$
$$v'_2 = 2d_2$$

and $F_{v'_1,v'_2,1-p_1}$ is the lower $100(1-p_1)$ percentage point of the F distribution with (v'_1,v'_2) degrees of freedom. This implies that

$$F_1 = \frac{T_1/v'_1}{T_2/v'_2}$$

$$= \frac{T_1/(d_1+1)}{T_2/d_2}$$

$$= F_{v'_1,v'_2,1-p_1} \qquad (2)$$

(ii) Similarly, if p_2 is such that

$$p_2 = \sum_{v=0}^{d_1-1} \binom{n}{v} \pi^v (1-\pi)^{n-v}$$

then

$$F_2 = \frac{T_1/v_1^*}{T_2/v_2^*}$$

$$= \frac{T_1/d_1}{T_2/(d_2+1)}$$

$$= F_{v_1^*,v_2^*,1-p_2} \qquad (3)$$

where

$$v_1^* = 2d_1$$

$$v_2^* = 2(d_2+1)$$

In other words it can be seen that, from equations (1) to (3), using the conditional binomial distribution $B(n,\pi)$ of d_1 is equivalent to "averaging" F_1 of equation (2) and F_2 of equation (3) (Le, 1991). The question is how to average these slightly different F statistics.

(i) One option is to take as degrees of freedom

$$v_1 = v_1^*$$
$$= 2d_1$$
$$v_2 = v_2'$$
$$= 2d_2$$

corresponding to the F statistic

$$\boxed{F = \frac{T_1/d_1}{T_2/d_2}}$$

as used by many (see Cox and Oakes, 1984, p. 84).

(ii) The other option is to average the degrees of freedom of F_1 and F_2, that is,

$$v_1 = (v_1' + v_1^*)/2$$
$$= 2d_1 + 1$$
$$v_2 = (v_2' + v_2^*)/2$$
$$= 2d_2 + 1$$

corresponding to the F statistic

$$F_c = \frac{T_1/(d_1 + .5)}{T_2/(d_2 + .5)}$$

as originally suggested by Cox (1953). There is empirical evidence to show that F performs slightly better than F_c with small samples. In fact,

$$F = \frac{\lambda_1}{\lambda_2} \frac{T_1/d_1}{T_2/d_2}$$

is distributed as $F(2d_1, 2d_2)$ (Cox and Oakes, 1984; in other words, $2\lambda_i T_i$ is treated as having a chi-square distribution with

$2d_i$ degrees of freedom—this result is exact without censoring). This makes it possible to construct confidence intervals for $\theta = \lambda_1/\lambda_2$.

Sample Size Determination

The determination of sample size is a crucial element in the design of a clinical trial or a survey. The planning of sample size can be approached in different ways, one of which is in terms of controlling the risk of making type I and type II errors. When a statistical test is performed, the probability of type I error, α, is chosen by the investigator or the statistician. However, the probability that a significant result would be obtained if a real difference between the two groups exists, or the power of the test $(1 - \beta;$ β being the probability of type II error) depends largely on the sample size. Thus one should take steps to ensure that the power of the clinical trial is sufficiently high to justify the effort involved. The customary procedure for the determination of the sample size, based on the test of significance approach, consists of the following four steps:

1. Specify the level of difference to be detected.
2. Specify the power of the experiment $(1 - \beta)$.
3. Specify necessary parameter(s); this usually involves some preliminary estimates or bounds for the population parameter(s).
4. The fourth and final step is a specification of allocation weights for determining group sizes, that is,

$$n_1 = nw_1$$
$$n_2 = nw_2$$
$$1 = w_1 + w_2$$

where n is the total size needed for the study. Perhaps the most common practice is setting $w_1 = w_2 = .5$ to have equal size subsamples.

The basic methods used for the comparison of survival distributions often are distribution-free or nonparametric in that no underlying assumptions about the distribution of time-to-event need be specified. For sample size determination, however, some such assumption must be made. The most common assumption is that survival time follows a constant-risk or exponential model, where $\lambda(t) = \lambda$. Under this model, we have

$$\sqrt{n}|\lambda_2 - \lambda_1| = (z_{1-\alpha} + z_{1-\beta})\sqrt{f(\lambda_1)/w_1 + f(\lambda_2)/w_2}$$

from which one can solve for total sample size n or power β through z_β. Here we have

λ_1, λ_2 = risks of populations 1 and 2, respectively

$$f(x) = x^2 \bigg/ \left[1 - \frac{e^{-x(\pi_2 - \pi_1)} - e^{-x\pi_2}}{x\pi_1}\right]$$

π_1, π_2 = recruiment period and total study lengths (see Chapter 1)

$z_{1-\alpha}, z_{1-\beta}$ = percentiles of the standard normal distribution

for example, $z_\alpha = 1.65$ for one-sided test with $\alpha = .05$. The logic can be seen as follows.

Consider the general family of statistics, say $\hat\theta$, that are normally distributed as $N(\theta_0, \Sigma_0^2)$ under a null hypothesis $\mathcal{H}_0$ and as $N(\theta_A, \Sigma_A^2)$ under an alternative hypothesis $\mathcal{H}_A$ where $\theta_A > \theta_0$ or $\theta_A < \theta_0$ and where Σ_0^2 and Σ_A^2 are some function of population variation measures and sample size. It is often true that $\Sigma_0^2 = \Sigma_A^2$. Given these distributions, one can then determine α and β (for type I and type II errors, respectively) as shown in Figure 3.2. From this figure we have

$$d_1 = z_{1-\alpha} \sum_0$$

$$d_2 = z_{1-\beta} \sum_A$$

and

$$|\theta_0 - \theta_A| = d_1 + d_2$$
$$= z_{1-\alpha} \sum_0 + z_{1-\beta} \sum_A$$

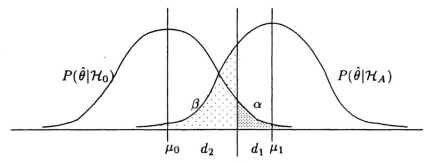

FIGURE 3.2. The distribution of a statistic $\hat{\theta}$ with variance Σ^2 under the null hypothesis $\mathcal{H}_0 : \mu = \mu_0$, i.e., the curve $P(\hat{\theta} \mid \mathcal{H}_0)$, and that under the alternative hypothesis $\mathcal{H}_A : \mu = \mu_A$ or the curve $P(\hat{\theta} \mid \mathcal{H}_A)$; and the probabilities of type I error (α) and type II error (β), where $X_0 = \mu_0 + z_\alpha \Sigma$.

where z_α is the $100(1-\alpha)$th percentile of the standard normal distribution. We now take

$$\theta = \lambda_2 - \lambda_1$$

so that under both $\mathcal{H}_0$ and $\mathcal{H}_A$

$$\Sigma^2 = \mathrm{Var}(\hat{\lambda}_2 - \hat{\lambda}_1)$$

where

$$\mathrm{Var}(\hat{\lambda}) = \lambda^2/nw \left[1 - \frac{e^{-\lambda(\pi_2 - \pi_1)} - e^{-\lambda \pi_2}}{\lambda \pi_1} \right]$$

as seen in Chapter 2.

Example 6. Consider that a clinical trial is to be conducted for a disease with moderate levels of mortality with a constant risk of $\lambda_1 = .3$ yielding 50 percent survivors after 2.3 years. Suppose that with treatment we are interested in a reduction in hazard to $\lambda_2 = .2$, that is, an increase in survival to 64 percent at 2.3 years. With equal-size groups,

$$w_1 = w_2 = .5$$

and

$$\alpha = .05$$

$$\beta = .10$$

$$z_{1-\alpha} = 1.645 \quad \text{for a one-sided test}$$

$$z_{1-\beta} = 1.283$$

If the study was to be terminated after $\pi_2 = 5$ years in which recruitment was to be limited to the first $\pi_1 = 3$ years, then

$$f(\lambda_1) = (.3)^2 \bigg/ \left[1 - \frac{e^{-(2)(.3)} - e^{-(5)(.3)}}{(.3)(3)}\right]$$

$$= .141$$

$$f(\lambda_2) = .081$$

therefore

$$\sqrt{n}(.1) = (1.65 + 1.28)\sqrt{(2)(.141 + .081)}$$

or

$$n = 382 \text{ subjects}$$

For practical purposes, the above formula for sample size determination should be used with certain modifications. The problem is that it may not be possible to find that many patients for the planned clinical trial. With the length of the recruitment period, π_1, fixed, it is often possible to estimate the number of patients available:

$$n = a\pi_1$$

where a is the estimated arrival rate (patients/year). Using this n, the researcher can either

(i) fix the follow-up period, $\pi_2 - \pi_1$, and evaluate β to see whether the effort is justifiable or
(ii) determine the follow-up time for a desirable level of β (this may not be feasible, for example, if the plan is too ambitious for the resources available).

Example 7. With $\pi_1 = 3$ years as in Example 4 but suppose that only $n = 300$ patients would be available ($a = 100$ patients/year):

(i) Using $\pi_2 = 5$ years, we have

$$\sqrt{300}(.1) = (1.65 + z_\beta)\sqrt{(2)(.141 + .081)}$$

or

$$z_{1-\beta} = .95$$

or a type II error of $\beta = .17$.

(ii) If we still want to maintain a $\beta = .10$, then it can be verified that a study of $\pi_2 = 6$ years is needed; or $\pi_2 - \pi_1 = 3$ years of follow-up.

Methods for Person-Years Data

The person-years method has been used extensively to analyze mortality in occupational health research and has been investigated by many authors; a brief description of these cohort studies was included in Section 2.3. Generally, we can subdivide the uses of the method into two main types of analysis:

1. One-sample analysis.
   For each individual in a cohort, define
   (a) Entry to the study
   (b) Potential follow-up time
   (c) The status

$$\delta_i = \begin{cases} 0 & \text{if censored} \\ 1 & \text{if died} \end{cases}$$

(d) An expected mortality, r_i, that is the cumulative risk that would have been expected if the subject had experienced similar death rates to those of the community of which the cohort under investigation is a part

For a cohort of size n, let

$$d = \sum \delta_i$$

$$e = \sum r_i$$

be the observed and expected numbers of deaths, respectively. The aim here is to investigate whether there is *excess* mortality.

2. Consider the case of two subcohorts with summarized figures (d_i, e_i); $i = 1, 2$. The aim is to *compare levels of excess mortality* of one subcohort versus the other.

Using the results for exponential models and those of Section 2.3, we suggest the following procedures for solving these two basic problems.

1. *For one-sample analysis:* Treat

$$2d\rho/\hat{\rho} = 2\rho e$$

as having a chi-square distribution with $2d$ degrees of freedom. This would allow us

(a) To construct confidence intervals for ρ
(b) To test for the null hypothesis

$$\mathcal{H}_0 : \rho = 1$$

Under $\mathcal{H}_0$, $2e$ has a chi-square distribution with $2d$ degrees of freedom.

2. *For the comparison between two subcohorts:* Treat

$$\theta \cdot \frac{e_2/d_2}{e_1/d_1}$$

as having an F distribution with $(2d_2, 2d_1)$ degrees of freedom where

$$\theta = \rho_2/\rho_1$$

This would allow us
(a) To construct confidence intervals for θ, the relative risk of subcohort 2 vs. subcohort 1
(b) To test for the null hypothesis

$$\mathcal{H}_0 : \theta = 1 \quad \text{(or } \rho_2 = \rho_1\text{)}$$

Under $\mathcal{H}_0$,

$$F = \frac{e_2/d_2}{e_1/d_1}$$

has an F distribution with $(2d_2, 2d_1)$ degrees of freedom (the value of this statistic serves as a point estimate for $1/\theta$; for more details, see Le, 1990).

Example 8. As an illustration, let us consider some data from Chapter 2, Example 7 (Table 2.8). Two groups of male asbestos workers with different levels of exposure were followed up for 10 years after initial exposure. Using gastrointestinal cancer as a specific cause of death, we have

Low-to-moderate exposure: $d_1 = 19$
$e_1 = 16.3$
Severe exposure: $d_2 = 39$
$e_2 = 17.7$

1. One-sample analysis:
   For the low-to-moderate exposure group, an application of the proposed method yields

   $$p = \Pr(\chi^2 \leq 32.6;\ df = 38)$$
   $$= .283$$

2. Comparison between two groups:
For the above two groups, severe versus low-to-moderate exposure, we have

$$F = \frac{17.7/39}{16.3/19}$$

$$= .53$$

at df = (78, 38), this F test yields $p = .009$ for a one-sided alternative or $p = .018$ for a two-sided alternative (the one-sided alternative is more obvious here).

Of course, we can apply the maximum-likelihood method of Section 2.3. However, by incorporating the sizes of the d's into its formulation, they are expected to work better with small samples. Being able to handle small samples would allow us to apply the person-years method to situations where

(a) The follow-up time is short resulting in small numbers of death d's.
(b) The cohort size is small-to-moderate (or even a sample from the cohort under investigation can be used).
(c) The specific cause of death is a rare disease.

3.3. PIECEWISE EXPONENTIAL: LINK BETWEEN PARAMETRIC AND NONPARAMETRIC METHODS

Consider the same two-sample problem with survival data

$$\{(t_{1i}, \delta_{1i})\} \quad i = 1, 2, \ldots, n_1$$

and

$$\{(t_{2j}, \delta_{2j})\} \quad j = 1, 2, \ldots, n_2$$

We pool the two samples together as in Section 4.1 and let

$$t_1 < t_2 < \cdots < t_m \qquad m \leq d \leq n_1 + n_2$$

be the distinct times with at least one event at each. In addition, let us suppose that the distributions have constant hazards between each pair $t_{i-1} < t_i$ of those distinct event times. For the interval $(t_{i-1}, t_i]$, let T_{li} and d_{li} be the total duration and number of events in group l; $l = 1, 2$. Since the distributions are now assumed exponential, recall from Section 3.2 that the comparison for interval $(t_{i-1}, t_i]$ is based on the conditional binomial distribution $B(d_i, \pi_i)$ where

$$d_i = d_{1i} + d_{2i}$$

and

$$\pi_i = T_{2i}/(T_{1i} + T_{2i})$$

Therefore, the evidence against the null hypothesis can be judged using the summarized standard normal z statistic:

$$z = \frac{\sum_{i=1}^{m} w_i(d_{2i} - d_i \pi_i)}{\sqrt{\sum_{i=1}^{m} d_i w_i^2 \pi_i (1 - \pi_i)}}$$

or z^2 is referred to as the chi-square distribution with 1 degree of freedom, where w_i is the weight associated with the interval $(t_{i-1}, t_i]$. The following are three possible choices:

(i) $w_i = 1$.
(ii) $w_i = n_i$, the total number of subjects at risk just before t_i.
(iii) $w_i = T_{1i} + T_{2i}$.

It can be seen that if all censored observations that occur in the interval (t_{i-1}, t_i) are adjusted to have occurred at t_{i-1}, then

1. The test with the above choice (i) is almost identical to the log-rank test.
2. Choice (ii) produces a test that is almost identical to the generalized Wilcoxon test.

The minor difference is due to the approximation of the hypergeometric variance in the denominator by the binomial variance. The following numerical example illustrates this equivalency.

Example 9. Refer back to the clinical trial to evaluate the effect of 6–MP to maintain remission from acute leukemia (Example 1). The performance of the log-rank can be summarized as in Table 3.3(a), which yields

$$\theta = \sum \{d_{2i} - E_0(d_{2i})\}$$
$$= 10.25$$

where

$$E_0(d_{2i}) = \frac{n_{2i}d_i}{n_i}$$
$$= d_i \Delta_i \quad \text{with} \quad \Delta_i = n_{2i}/n_i$$

and

$$\text{Var}_0(\theta) = \sum \text{Var}_0(d_{2i})$$
$$= \sum \frac{n_{1i}n_{2i}d_i(n_i - d_i)}{n_i^2(n_i - 1)}$$
$$= \frac{d_i(n_i - d_i)}{n_i - 1}\Delta_i(1 - \Delta_i)$$
$$= 6.2570$$

TABLE 3.3. Comparison of Log-Rank and Piecewise Exponential Procedures

(a) *Log-Rank Procedure*

Event Time		At risk				
(2) 6–MP	(1) Control	(2)	(1)	Total	d_i	$\Delta_i = n_{1i}/n_i$
23	23	6	1	7	2	.1429
22	22			9	2	.2222
—	17	10	3	13	1	.2308
16	—	11	3	13	1	.2308
—	15	11	4	15	1	.2667
13	—	12	4	16	1	.2500
—	12,12	12	6	18	2	.3333
—	11,11	13	8	21	2	.3810
10	—	15	8	23	1	.3478
—	8,8,8,8	16	12	28	4	.4286
7	—	17	12	29	1	.4138
6,6,6	—	21	12	23	3	.3636
—	5,5	21	16	35	2	.4000
—	4,4	21	16	37	2	.4324
—	3	21	17	38	1	.4474
—	2,2	21	19	40	2	.4750
—	1,1	21	21	42	2	.5000

(b) *Piecewise Exponential Procedure*

		Time from Previous Event				
(2)	(1)	T_{2i}	T_{2i}	Total	d_i	$\pi_i = T_{1i}/(T_{1i} + T_{2i})$
23	23	6	1	7	2	.1429
22	22	40	10	50	2	.2000*
—	17	10	3	13	1	.2308
16	—	11	3	14	1	.2143
—	15	22	8	30	1	.2667
13	—	12	4	16	1	.2500
—	12,12	12	6	18	2	.3333
—	11,11	13	8	21	2	.3810
10	—	31	16	47	1	.3404*
—	8,8,8,8	16	12	28	4	.4286
7	—	17	12	29	1	.4138
6,6,6	—	21	12	33	3	.3636
—	5,5	21	14	35	2	.4000
—	4,4	21	16	37	2	.4324
—	3	21	17	38	1	.4474
—	2,2	21	19	40	2	.4750
—	1,1	21	21	42	2	.5000

These results lead to

$$\chi^2 = \frac{(10.25)^2}{6.2570}$$

$$= 16.75$$

The performance of the piecewise exponential procedure is summarized in Table 3.3(b), from which we have

$$\chi^2 = \frac{\left[\sum(d_{2i} - d_i\pi_i)\right]^2}{\sum d_i\pi_i(1-\pi_i)}$$

$$= \frac{(10.302)^2}{6.5678}$$

$$= 16.16$$

If we adjust censored observations in intervals $(8,10)$ and $(17,22)$, the two chi-squared statistics would be almost the same.

3.4. TRENDS IN SURVIVAL

In the comparison of several treatments, the main question is whether there is any difference among the treatments. To answer that question, one may wish to test the null hypothesis of no difference. However, in deciding on an appropriate testing procedure, it is not enough to specify the null hypothesis $\mathcal{H}_0$. One must also be clear about the nature of the alternatives against which $\mathcal{H}_0$ is being tested. Suppose now that the treatments are ordered (of course, before survival times have been obtained) in such a way that under the alternative one could expect better survival under treatment 2 than under treatment 1, under treatment 3 than under treatment 2, and so on. One example is the case where we administer a chemical treatment at k doses to k groups of subjects; group 1 is placebo. The use of the above chi-square test with $(k-1)$ degrees of freedom is no longer appropriate (or powerful enough) since it rejects $\mathcal{H}_0$ whenever the difference between any two groups is sufficiently large, regardless of their order, while in the present case only a "trend" in survival would support the alternative over the null hypothesis.

Class of Tests Against Stochastically Ordered Alternatives

Analyses of continuously measured covariates often assume a proportional hazards model (Cox, 1972). An ordinal covariate does not have a defined metric, so linearity of proportional hazards model is not totally meaningful. However, the inherent ordering of categories allows consideration of monotonicity and a test of stochastically ordered distributions could be formed as follows (Huang, 1994).

Let $S_i(t)$, $i = 1,\ldots,R$, be survival functions for the R groups defined by the value of an ordinal risk factor. For all t,

$$\mathcal{H}_0 : S_1(t) = S_2(t) = \cdots = S_i(t) = \cdots = S_R(t)$$

$$\mathcal{H}_A : S_1(t) \geq S_2(t) \geq \cdots \geq S_i(t) \geq \cdots \geq S_R(t)$$

and for at least one i, $S_i(t) > S_{i+1}(t)$

Suppose in the combined sample of R groups, we observe events at K distinct time points

$$t_1 < t_2 < \cdots < t_k < \cdots < t_K$$

At each t_k, we form a $2 \times R$ table as follows:

Group	1	2	$\cdots$	i	$\cdots$	j	$\cdots$	R	Total
Deaths	d_{1k}	d_{2k}	$\cdots$	d_{ik}	$\cdots$	d_{jk}	$\cdots$	d_{Rk}	d_k
Survivors	s_{1k}	s_{2k}	$\cdots$	s_{ik}	$\cdots$	s_{jk}	$\cdots$	s_{Rk}	s_k
At risk	n_{1k}	n_{2k}	$\cdots$	n_{ik}	$\cdots$	n_{jk}	$\cdots$	n_{Rk}	n_k

Here, n_{ik} is the number of individuals at risk in the ith group at t_k^-, d_{ik} is the number of events at t_k, and $s_{ik} = n_{ik} - d_{ik}$.

A measure of ordinal association for this $2 \times R$ table could be expressed as

$$\theta_{wk} = \sum_{i<j, n_{ik}+n_{jk}\neq 0} w_{ijk} \left[d_{jk} - \frac{n_{jk}}{n_{ik}+n_{jk}}(d_{ik}+d_{jk}) \right]$$

$$= \sum_{i<j, n_{ik}+n_{jk}\neq 0} w_{ijk} \left[\frac{n_{ik}}{n_{ik}+n_{jk}} d_{jk} - \frac{n_{jk}}{n_{ik}+n_{jk}} d_{ik} \right]$$

$$= \sum_{i=1}^{R} d_{ik} \left[\sum_{j\neq i, n_{ik}+n_{jk}\neq 0} \frac{w_{ijk} n_{jk}}{n_{ik}+n_{jk}} \operatorname{sgn}(i-j) \right]$$

where sgn( ) is the sign function and w_{ijk} is a certain kind of weight associated with groups i and j, and t_k.

Define

$$T_{wik} \equiv \sum_{j\neq i, n_{ik}+n_{jk}\neq 0} \frac{w_{ijk} n_{jk}}{n_{ik}+n_{jk}} \operatorname{sgn}(i-j)$$

which is determined by marginals and weight; the above expression could be rewritten as

$$\theta_{wk} = \sum_{i=1}^{R} d_{ik} T_{wik} \qquad (4)$$

Under the $\mathcal{H}_0$:

$$\mathrm{E}_0[\theta_{wk}] = 0 \qquad (5)$$

$$\mathrm{Var}_0(\theta_{wk}) = \frac{d_k s_k}{n_k(n_k-1)} \sum_{i=1}^{R} n_{ik} T_{wik}^2 \qquad (6)$$

The derivation of the variance under $\mathcal{H}_0$ is as follows: Given marginals in the $2 \times R$ table, d_{ik}, d_{jk}, and $d_{ik}+d_{jk}$ are all hypergeometrically distributed under $\mathcal{H}_0$:

$$d_{ik} \sim \text{Hyper}(n_k, n_{ik}, d_k)$$
$$d_{jk} \sim \text{Hyper}(n_k, n_{jk}, d_k)$$
$$d_{ik} + d_{jk} \sim \text{Hyper}(n_k, n_{ik} + n_{jk}, d_k)$$

We have

$$\text{Var}_0(d_{ik}) = \frac{d_k s_k n_{ik}(n_k - n_{ik})}{n_k^2(n_k - 1)}$$

$$\text{Var}_0(d_{jk}) = \frac{d_k s_k n_{jk}(n_k - n_{jk})}{n_k^2(n_k - 1)}$$

$$\text{Var}_0(d_{ik} + d_{jk}) = \frac{d_k s_k (n_{ik} + d_{jk})(n_k - n_{ik} - n_{jk})}{n_k^2(n_k - 1)}$$

$$\text{Cov}_0(d_{ik}, d_{jk}) = \tfrac{1}{2}[\text{Var}_0(d_{ik} + d_{jk}) - \text{Var}_0(d_{ik}) - \text{Var}_0(d_{jk})]$$
$$= -\frac{d_k s_k n_{ik} n_{jk}}{n_k^2(n_k - 1)}$$

So,

$$\text{Var}_0(\theta_{wk}) = \text{Var}_0\left(\sum_{i=1}^{R} d_{ik} T_{wik}\right)$$

$$= \sum_{i=1}^{R} T_{wik}^2 \text{Var}_0(d_{ik}) + \sum_{i \neq j} T_{wik} T_{wjk} \text{Cov}_0(d_{ik}, d_{jk})$$

$$= \frac{d_k s_k}{n_k^2(n_k - 1)} \left[\sum_{i=1}^{R} n_{ik}(n_k - n_{ik}) T_{wik}^2 - \sum_{i \neq j} n_{ik} n_{jk} T_{wik} T_{wjk}\right]$$

$$= \frac{d_k s_k}{n_k(n_k - 1)} \sum_{i=1}^{R} n_{ik} T_{wik}^2 - \frac{d_k s_k}{n_k^2(n_k - 1)} \left(\sum_{i=1}^{R} n_{ik} T_{wik}\right)^2$$

$$= \frac{d_k s_k}{n_k(n_k - 1)} \sum_{i=1}^{R} n_{ik} T_{wik}^2$$

A class of tests against $\mathcal{H}_A$ employing the above statistic can be formed as follows:

$$\theta_w = \sum_{k=1}^{K} \theta_{wk}$$

$$= \sum_{k=1}^{K} \sum_{i=1}^{R} d_{ik} T_{wik} \qquad (7)$$

The null hypothesis is rejected if θ_w or $|\theta_w|$ is large; θ_w is asymptotically normally distributed under $\mathcal{H}_0$ with mean zero, and its variance is consistently estimated by

$$\text{Var}_0^{\text{est}}(\theta_w) = \sum_{k=1}^{K} \text{Var}_0(\theta_{wk})$$

$$= \sum_{k=1}^{K} \frac{d_k s_k}{n_k(n_k - 1)} \sum_{i=1}^{R} n_{ik} T_{wik}^2 \qquad (8)$$

Decision is based on the standardized statistic

$$z = \theta_w / \{\text{Var}_0^{\text{est}}(\theta_w)\}^{1/2}. \qquad (9)$$

and referred to the standard normal percentile $z_{1-\alpha}$ for a specified size α of the test.

Weight Choices

This class of tests consists of members with various weights. As seen from derivation, a valid weight, w_{ijk}, should only use information available at t_k^- and

$$w_{ijk} = w_{jik} \qquad (10)$$

In a specific situation, one may be more powerful than others. Some potential weight choices could be borrowed from those of the Tarone and Ware (1977) class statistic since θ_w is expressible as summation of Tarone–Ware statistics for comparing every pair of groups:

$$\theta_w = \sum_{k=1}^{K} \theta_{wk}$$

$$= \sum_{k=1}^{K} \sum_{i<j, n_{ik}+n_{jk}\neq 0} w_{ijk} \left[d_{jk} - \frac{n_{jk}}{n_{ik}+n_{jk}}(d_{ik}+d_{jk}) \right]$$

$$= \sum_{i<j} \sum_{k=1, n_{ik}+n_{jk}\neq 0}^{K} w_{ijk} \left[d_{jk} - \frac{n_{jk}}{n_{ik}+n_{jk}}(d_{ik}+d_{jk}) \right] \qquad (11)$$

As a consequence, some popular weight choices below for Tarone–Ware class test may be considered for w_{ijk}.

The above class of nonparametric tests against stochastically ordered alternatives for censored survival data could be viewed as an extension of Tarone–Ware class statistic to R-group situation. The primary aim is for testing the independence between censored survival time and an ordinal covariate. This class of tests includes the Le–Grambsch–Louis method (1994) and Tarone's test for survival trend as special cases as shown in the following sections. In addition, Le et al. proved that the Le–Grambsch–Louis method includes a generalization of Jonckheere's (1954); also O'Brien's rank procedure (1978) for ordinal covariate was shown to be a special case of the Le–Grambsch–Louis method without ties.

Tarone Test

In this proposed class of tests, some available score x_i assigned to the ith group may also be built into the weight choice—for example, $x_i = \log$(dose administered to group i) in a drug dose study case. Tarone (1975) proposed such a test against trend, which will be shown as a special case in our proposed class as follows.

Taking $w_{ijk} = (n_{ik} + n_{jk})(x_i - x_j)\text{sgn}(i-j)/n_k$,

$$T_{wik} = \sum_{j \neq i, n_{ik}+n_{jk} \neq 0} \frac{w_{ijk} n_{jk}}{n_{ik} + n_{jk}} \text{sgn}(i-j)$$

$$= \sum_{j \neq i} \frac{(x_i - x_j) n_{jk}}{n_k}$$

$$\theta_w = \sum_{k=1}^{K} \theta_{wk}$$

$$= \sum_{k=1}^{K} \sum_{i=1}^{R} d_{ik} T_{wik}$$

$$= \sum_{k=1}^{K} \sum_{i=1}^{R} \sum_{j \neq i} \frac{d_{ik}(x_i - x_j) n_{jk}}{n_k}$$

$$= \sum_{k=1}^{K} \sum_{i=1}^{R} x_i \left[d_{ik} - \frac{d_k n_{ik}}{n_k} \right]$$

which is the statistic for Tarone's score test. Consequently our variance estimate would be reduced to that of Tarone's. It should be noted that the term in the above sum is similar to the "sum of products" in regression and correlation analysis, the correlation between the assigned score and the deviation from $\mathcal{H}_0$.

Le–Gramsch–Louis Method

Consider the following three examples:

(i) A comparison of four smoking groups: nonsmokers, ex-smokers, current smokers of low-tar cigarettes, and current smokers of high-tar cigarettes

(ii) A comparison of cancer patients in three stages, I, II, and III; higher stages are more advanced

(iii) A comparison of three diets: low-fat, saturated fat, and unsaturated fat

These examples have a feature in common: The various groups or stages follow natural ordering and can be considered as various categories or levels of an ordinal covariate. Of course, we still can apply the Tarone test to these problems; however, such an application depends on the availability of the scores x_j's. Le, Grambsch, and Louis (1994) proposed a procedure where such scores are not needed. The Le–Grambsch–Louis test statistic against trend is based on number of concordances and discordances of the $2 \times R$ table at each event time point, which could be shown as a special case in this proposed class of tests. A proof is as follows.

Setting $w_{ijk} = (n_{ik} + n_{jk})w_k$, where w_k is weight associated with t_k, we have

$$T_{wik} = \sum_{j \neq i, n_{ik}+n_{jk} \neq 0} \frac{w_{ijk} n_{jk}}{n_{ik} + n_{jk}} \operatorname{sgn}(i-j)$$

$$= \sum_{j \neq i} w_k n_{jk} \operatorname{sgn}(i-j)$$

$$\theta_w = \sum_{k=1}^{K} \theta_{wk}$$

$$= \sum_{k=1}^{K} \sum_{i=1}^{R} d_{ik} T_{wik}$$

$$= \sum_{k=1}^{K} w_k \sum_{i=1}^{R} \sum_{j \neq i} d_{ik} n_{jk} \operatorname{sgn}(i-j)$$

$$= \sum_{k=1}^{K} w_k \left[\sum_{j=1}^{R-1} s_{jk} \sum_{i=j+1}^{R} d_{ik} - \sum_{i=1}^{R-1} d_{ik} \sum_{j=1+1}^{R} s_{jk} \right]$$

$$\operatorname{Var}_0^{\operatorname{est}}(\theta_w) = \sum_{k=1}^{K} \operatorname{Var}_0(\theta_{wk})$$

$$= \sum_{k=1}^{K} \frac{d_k s_k}{n_k(n_k-1)} \sum_{i=1}^{R} n_{ik} T_{wik}^2$$

$$= \sum_{k=1}^{K} w_k^2 \frac{d_k s_k}{n_k(n_k-1)} \sum_{i=1}^{R} n_{ik} \left[\sum_{j \neq i} n_{jk} \operatorname{sgn}(i-j) \right]^2$$

$$= \sum_{k=1}^{K} w_k^2 \frac{d_k s_k \left(n_k^3 - \sum_{i=1}^{R} n_{ik}^3 \right)}{3 n_k(n_k-1)}$$

Both the test statistic and the variance estimate are reduced to those of the Le–Grambsch–Louis method. In this formulation,

$$C_k = \sum_{j=1}^{R-1} s_{jk} \left[\sum_{i=j+1}^{R} d_{ik} \right]$$

and

$$D_k = \sum_{i=1}^{R-1} d_{ik} \left[\sum_{j=i+1}^{R} s_{jk} \right]$$

are the number of concordant pairs and the number of discordant pairs in the $2 \times R$ contingency table at time t_k (see Chapter 1).

Example 10. Consider an experiment or an observational study in which we have R groups with some natural ordering so that they can be considered as R categories of an ordinal covariate X. For example, a laboratory investigator interested in the relation between diet and the development of tumors divided 90 rats into 3 groups and fed them with low-fat, saturated-fat, and unsaturated-fat diets, respectively (King et al., 1979; Lee, 1980). The rats were of the same age and species and were in similar conditions. An identical number of tumor cells were injected into a foot pad of each rat and the rats were observed for 200 days. Many rats developed a recognizable tumor early, some were tumor-free at the end, and two died before study termination but without evidence of tumor. Data are presented in Table 3.4. An application of the above methods yields

$$z = 3.26$$

TABLE 3.4. Tumor-Free Time (in Days) of 90 Rats on Three Different Diets

Rat	Low-fat	Rat	Saturated	Rat	Unsaturated
1	140	1	124	1	112
2	177	2	58	2	68
3	50	3	56	3	84
4	65	4	68	4	109
5	86	5	79	5	153
6	153	6	89	6	143
7	181	7	107	7	60
8	191	8	86	8	70
9	77	9	142	9	98
10	84	10	110	10	164
11	87	11	96	11	63
12	56	12	142	12	63
13	66	13	86	13	77
14	73	14	75	14	91
15	119	15	117	15	91
16	140$^+$	16	98	16	66
17	200$^+$	17	105	17	70
19	200$^+$	19	43	19	63
20	200$^+$	20	46	20	66
21	200$^+$	21	81	21	66
22	200$^+$	22	133	22	94
23	200$^+$	23	165	23	101
24	200$^+$	24	170$^+$	24	105
25	200$^+$	25	200$^+$	25	108
26	200$^+$	26	200$^+$	26	112
27	200$^+$	27	200$^+$	27	112
28	200$^+$	28	200$^+$	28	126
29	200$^+$	29	200$^+$	29	161
30	200$^+$	30	200$^+$	30	178

Data taken from Lee, 1980.

(weights are chosen to be unity for simplicity). This indicates a highly significant trend in survival.

It can be seen that if the categories are defined by a qualitative classification or an ordinal covariate as in our numerical example, any choice of a system of scoring is arbitrary and may be difficult to interpret. In the absence of an obvious system of scoring, people often use equally spaced scores. However, in many cases there may be little virtue in using equally spaced scores or doses; then, this rank correlation-based method is perhaps more objective. The test can be considered a generalization of Jonckheere's test with $w_k = 1$.

Similarly, if we choose the weights to be $w_k = 1/n_k$, then the test statistic can be written as

$$\theta_{1/n} = \sum_{j<m} U_{jm}$$

where U_{jm} is the log-rank statistic for the comparison of group j versus group m ($i \leq j < m \leq R$).

Note: An SAS program for the Tarone's score test would include these instructions: PROC LIFETEST METHOD=KM; TIME WEEKS*FAILURE(0); TEST MAKEUP; where KM stands for Kaplan–Meier method, WEEKS is the variable name for duration time, FAILURE the variable name for survival status, "0" is the coding for censoring, and MAKEUP is the variable name specifying scores (these scores are assigned to individuals, same score for all individuals in the same group). No packaged programs are available for other trend tests.

Test for Trend in Constant Hazards and Its Application in Occupational Health

Recall that in the person-years method, for each member of a well-defined cohort, we have

(i) Entry to the study at time 0
(ii) Follow-up to time t_i
(iii) Death at time T_i (censored if $T_i > t_i$). Let

$$\delta_i = \begin{cases} 0 & \text{if censored} \\ 1 & \text{if died} \end{cases}$$

(iv) An expected mortality (m_i) that is the cumulative risk that would have been expected if the subject had experienced similar death rates to those of the referenced population.

For a cohort of size n, let

$$d = \sum \delta_i$$

$$m = \sum m_i$$

be the observed and expected deaths, respectively. The ratio

$$\text{SMR} = d/m$$

is called the *standardized mortality ratio* (SMR) and is used as a measure of excess mortality. As an example, consider a cohort of some 7000 British workers exposed for several years to vinyl chloride monomer; this cohort was followed for several years to determine whether their mortality experience differed from that of the British general population (Example 9 and Table 2.10, Chapter 2). Data, reproduced in the following table, are for deaths from cancers and are tabulated separately for four groups based on years since entering the industry.

Deaths from Cancers	Years Since Entering the Industry				
	1–4	5–9	10–14	15+	
Observed	9	15	23	68	115
Expected	20.3	21.3	24.5	60.8	126.8
SMR (%)	44.5	70.6	94.0	111.8	90.7

The above data display shows some interesting features commonly seen in studies of occupational health:

(i) For the group with 1–4 years since entering the industry, we have a death rate that is less than half of the referenced population (SMR = 44.5 percent). This phenomenon, known as the "healthy worker" effect, is a consequence of a selection factor whereby workers are necessarily in good health at the time of their entry into the work force.

(ii) We see an attenuation of the healthy worker effect with the passage of time, so that the cancer death rates show a slight excess after 15 years (SMR = 111.8 percent). To prove the attenuation of the healthy worker effect with the passage of time is to prove the increase of occupational mortality for the four groups in the above table; that is, a trend in survival. Of course, a possible procedure is to allot a score x_i to the ith subcohort and to perform some sort of regression analysis. However, the Tarone test is less efficient in this case because it takes no account of the particular models for person-years data.

Recall also, from Chapter 2, that if we assume the proportional hazards model

$$\lambda(t) = \rho \lambda_0(t)$$

where $\lambda(t)$ is the hazard function for the cohort, $\lambda_0(t)$ is the hazard function for the referenced population, and ρ is the relative risk of the cohort [of which the SMR is the maximum-likelihood estimate (MLE)], then $\{(m_i, \delta_i)\}$ can be treated as a censored sample from an exponential distribution with hazard parameter ρ.

In general, let us suppose there are k subcohorts. For the ith member of the jth subcohort, let t_{ji}, δ_{ji}, and ρ_i be the exposure time, status, and relative risk, respectively. Also let m_{ji} be the cumulative risk and

$$m_j = \sum_i m_{ji} \qquad m = \sum m_j$$

$$d_j = \sum_i d_{ji} \qquad d = \sum d_j$$

Let x_j be the score associated with the jth cohort. If we represent the relative risk ρ_j by the linear-trend model

$$\rho_j = \exp(\beta_1 + \beta_2 x_j)$$

then the likelihood function is expressible by

$$L = \prod_j \left\{ \prod_i [\exp(\beta_1 + \beta_2 x_j)]^{\delta_{ji}} \exp[-m_{ji} e^{\beta_1 + \beta_2 x_j}] \right\}$$
$$= \prod_j \exp[d_j(\beta_1 + \beta_2 x_j)] \exp[-m_j e^{\beta_1 + \beta_2 x_j}]$$

or

$$\ln L = \sum_j \{d_j(\beta_1 + \beta_2 x_j) - m_j \exp(\beta_1 + \beta_2 x_j)\}$$

and the null hypothesis of no trend is expressible by

$$\mathcal{H}_0 : \beta_2 = 0$$

Under $\mathcal{H}_0$, β_1 is estimated by

$$\hat{\beta}_1 = \ln(d/m)$$

Therefore,

$$\frac{\partial \ln L}{\partial \beta_1} = \sum_j \{d_j - m_j \exp(\beta_1 + \beta_2 x_j)\}$$
$$\stackrel{\hat{}}{\underset{\mathcal{H}_0}{=}} 0$$

$$\frac{\partial \ln L}{\partial \beta_2} = \sum_j \{d_j x_j - m_j x_j \exp(\beta_1 + \beta_2 x_j)\}$$
$$\stackrel{\hat{}}{\underset{\mathcal{H}_0}{=}} \sum_j \{d_j x_j - m_j x_j d/m\}$$
$$= d(\mu_d - \mu_m)$$

where

$$\mu_d = \left(\sum d_j x_j\right) \bigg/ \left(\sum d_j\right)$$

and

$$\mu_m = \left(\sum m_j x_j\right) \Big/ \left(\sum m_j\right)$$

are the weighted means of the scores using observed numbers of deaths (d_j's) and expected number of deaths (m_j's) as weights, respectively. We also have

$$-\frac{\partial^2 \ln L}{\partial \beta_1^2} = \sum_j m_j \exp(\beta_1 + \beta_2 x_j)$$

$$\underset{\mathcal{H}_0}{\hat{=}} d$$

$$-\frac{\partial^2 \ln L}{\partial \beta_1 \partial \beta_2} = \sum_j m_j x_j \exp(\beta_1 + \beta_2 x_j)$$

$$\underset{\mathcal{H}_0}{\hat{=}} d(d/m) \sum m_j x_j$$

$$-\frac{\partial^2 \ln L}{\partial \beta_2^2} = \sum_j m_j x_j^2 \exp(\beta_1 + \beta_2 x_j)$$

$$\underset{\mathcal{H}_0}{\hat{=}} (d/m) \sum m_j x_j^2$$

From these results we have

$$X_{ES}^2 = [0 \quad d(\mu_d - \mu_m)] \begin{bmatrix} d & (d/m)\sum m_j x_j \\ (d/m)\sum m_j x_j & (d/m)\sum m_j x_j^2 \end{bmatrix}^{-1}$$

$$\times \begin{bmatrix} 0 \\ d(\mu_d - \mu_m) \end{bmatrix}$$

$$= \frac{1}{\sigma_m^2} [0 \quad d(\mu_d - \mu_m)] \begin{bmatrix} \dfrac{\sum m_j x_j^2}{m} & -d\mu_m \\ d\mu_m & \dfrac{1}{d} \end{bmatrix} \begin{bmatrix} 0 \\ d(\mu_d - \mu_m) \end{bmatrix}$$

$$= \frac{d(\mu_d - \mu_m)^2}{\sigma_m^2}$$

where

$$\sigma_m^2 = \left[\sum m_j(x_j - \mu_m)^2\right] \Big/ \left(\sum m_j\right)$$

is the weighted variance of the scores x_j's using the expected numbers of deaths as weights. Taking the square root we have

$$z = \frac{d^{1/2}(\mu_d - \mu_m)}{\sigma_m}$$

and the value of the z score may be referred to the standard normal percentiles for a statistical decision (Le, 1993). Of course, the calculations cannot be performed until the scores x_j's have been chosen. If the subcohorts are defined by a qualitative classification (e.g., race), we may choose equally spaced x_j's. On the other hand, if the cohorts are defined by a measurement (e.g., age or length of exposure), it is more reasonable to choose the scores related to the values assumed by the measurement (e.g., taking midpoints of the groups if possible). Even the choice of the scores is often arbitrary, the presence of the weighted variance/standard deviation (σ_m^2 or σ_m) in the denominator of X_{ES}^2/z would mostly neutralize the effects of arbitrary choices; different choices will usually give fairly close results.

Example 11. Refer to the data of Example 8 (or the table of this section) and suppose we use the score

$$x_j = j \qquad j = 1, 2, 3, \text{ and } 4$$

Then we have

$$\mu_d = 3.30$$
$$\mu_m = 2.99$$
$$\sigma_m^2 = 1.29$$

leading to

$$z = 2.96 \qquad (p = .003)$$

3.5. COMPARISON OF TWO CLUSTERED SAMPLES

In this section, a study of otitis media is described that requires a test for the comparison of two clustered samples of censored data. A special method is formulated taking into account the within-subject correlation in the formation of the log-rank statistic (Le and Lindgren, 1996).

Problem

Inflammation of the middle ear, or otitis media (OM), is one of the most common childhood illnesses and accounts for one-third of the practice of pediatrics during the first 5 years of life; the most common form is otitis media with effusion (OME). Treatments of otitis media are of considerable importance due to morbidity for the child as well as concern about long-term effects on behavior, speech, and language development. When frustrated with failure of medical therapy, patients and physicians turn to surgical intervention; 1 million children are estimated to receive ventilating tubes each year in the United States. Tubes have been shown to decrease the incidence of otitis media episodes and improve hearing as long as they are in place and functioning.

A study enrolled 78 children age 6 months to 8 years with chronic OME (at least 56 days of documented OME during the 90 days preceding enrollment) who were having therapeutic myringotomy for bilateral tympanostomy tube placement (intubation). Eligible children were included if they had received all of their preceding health care from staff pediatricians, they had current immunization status, and their parents gave signed informed consent. Eligible children were excluded if they could not return for regular follow-up, had sensorineural hearing loss, a complicating illness, infection, chronic disease, structural middle ear damage, prior middle ear surgery other than intubation for OME, prior adenoidectomy or tonsillectomy, intracranial complications of OM, chronic rhinitis or sinusitis requiring regular decongestant or antihistamine medications, or congenital malformations of the external, middle, or inner ear. Subjects were examined before and two weeks after surgery, and every 3 months thereafter. Investigators randomized 40 children to receive 2-week trials of prednisone and sulfamethoprim treatment immediately after surgery. The remaining 38 had no additional intervention after

COMPARISON OF TWO CLUSTERED SAMPLES

surgery and served as controls. The aim of the medical treatment is to prolong the life of the tubes, the primary endpoint being the cessation of tube functioning (blocked) or tube extrusion.

In order to determine possible effects of medical treatment in the prolonging of tube life, we can just compare the two samples of censored tube survival times: ears with treatment vs. controls. However, a straightforward application of a procedure such as the log-rank test is not appropriate because observations are not independent; with tubes in both left and right ears of the same child. Therefore, it is desirable to have a test for comparing two such clustered samples of censored data; the actual data are listed in Appendix C.

Data Summarization

Suppose that in the combined sample of all children we observe events—tubes extruded or blocked—at k distinct time points, $t_1 < t_2 < \cdots < t_i < \cdots < t_k$, where $t = 0$ is the time origin when a tube was surgically inserted. At the t_i, $1 \leq i \leq k$, there is a set R_i of ears at risk, consisting of those ears that have been followed for at least t_i units of time. The risk set R_i can be decomposed into two parts:

(i) R_{1i} is the subset of R_i where each child has only one ear's tube (in either left or right ear) at risk at time t_i^-; these children entered the study with tubes in both ears but one of the two tubes had previously (before t_i) become extruded or blocked.

(ii) R_{2i} is the subset of R_i where each child has both ears' tubes at risk at time t_i^-.

We arrange the data in R_{1i} into a 2×2 table (Table 3.5) as in the formation of the log-rank statistic (a plus sign in a subscript indicates summation over that subscript). For entries in Table 2.6, a_{jmi}, the first subscript j is the group indicator (1 = medical, 0 = control), the second subscript m is the event indicator (1 = event at t_i, 0 = no event or "survivor"), and the last subscript i denotes time ($i = 1, 2, \ldots, k$).

Similarly, we arrange the "ears" in R_{2i} as shown in Table 3.6. However, in this group, each child had two ears at risk; an alternative tabulation is to count the numbers of "children" and arrange them into a 2×3 table as shown in Table 3.7. The second subscript m in c_{jmi} denotes the number of events (2, 1, or 0) for a child in R_{2i}; as

TABLE 3.5. Arrangement of Children with One Ear's Tube at Risk

Treatment Group	Event at t_i		Totals
	Yes	No	
Medical	a_{11i}	a_{10i}	a_{1+i}
Control	a_{01i}	a_{00i}	a_{0+i}
Totals	a_{+1i}	a_{+0i}	$a_{++i} = n_{1i}$

TABLE 3.6. Arrangement of Ear Tubes for Children with Two Tubes at Risk

Treatment Group	Event at t_i		Totals
	Yes	No	
Medical	b_{11i}	b_{10i}	b_{1+i}
Control	b_{01i}	b_{00i}	b_{o+i}
Totals	b_{+1i}	b_{+0i}	b_{++i}

TABLE 3.7. Arrangement of Children with Two Ears' Tubes at Risk

Treatment Group	Events at Time t_i			Totals
	Both	One ear	None	
Medical	c_{12i}	c_{11i}	c_{10i}	c_{1+i}
Control	c_{02i}	c_{01i}	c_{00i}	c_{0+i}
	c_{+2i}	c_{+1i}	c_{+0i}	$c_{++i} = n_{2i}$

compared to Table 3.6, we have

$$b_{j1i} = 2c_{j2i} + c_{j1i} \quad j = 1, 2$$
$$b_{+1i} = 2c_{+2i} + c_{+1i}$$
$$b_{1+i} = 2c_{1+i}$$
$$b_{++i} = 2n_{2i}$$

Models

Model and a Statistic from R_{1i}

R_{1i} is the subset of the risk set R_i where each child has only one ear's tube (in either left or right ear) at risk at time t_i^-; these children entered the study with tubes in both ears, but one of the two tubes had previously (before t_i) become extruded or blocked. Since the failure of the other ear may very well affect the hazard rate of the ear now at risk, the usual proportional hazards model is now modified as follows; for the control group,

$$\Pr(T = t_i \mid T \geq t_i) = \eta \lambda_i$$

and for the medical group,

$$\Pr(T = t_i \mid T \geq t_i) = \rho \eta \lambda_i$$

where η represents the effect of the failed ear and ρ the treatment effect.

Considering the statistic a_{11i} in Table 3.1, we have under the null hypothesis, $H_0 : \rho = 0$, of no medical treatment effect

$$E_0(a_{11i}) = \frac{a_{+1i} a_{1+i}}{a_{++i}}$$

and its variance is estimated by (hypergeometric model)

$$\widehat{\mathrm{Var}}_0(a_{11i}) = \frac{a_{+1i} a_{+0i} a_{1+i} a_{0+i}}{a_{++i}^2 (a_{++i} - 1)}$$

The constant η was cancelled due to the multiplicative nature of the model. For example,

$$E(a_{11i}) = (a_{1+i})(\eta \lambda_i)$$

and under the null hypothesis of no treatment effects, $\eta\lambda_i$ is estimated by $(a_{+1i})/(a_{++i})$ leading to:

$$E_0(a_{11i}) = \frac{a_{+1i}a_{1+i}}{a_{++i}}$$

Model and a Statistic from R_{2i}

R_{2i} is the subset of the risk set R_i where each child has both ears' tubes at risk at time t_i^-. We adopt Rosner's model (Rosner, 1982) and express as follows: Similarly, if we consider statistic b_{11i} of Table 3.6, then

$$E_0(b_{11i}) = \frac{b_{+1i}b_{1+i}}{b_{++i}}$$

In order to estimate $\text{Var}(b_{11i})$, we adopt Rosner's model and method for binary paired data (Rosner, 1982) and proceed as follows:

For a particular group, say *medical group*, let T_{jk} be the survival time of the kth ear ($k = 1$ or 2) of the jth child. Let λ_i be the instantaneous event rate (i.e., hazard) for the medical group at time t_i; the adopted Rosner's model is expressed as:

$$\lambda_i = \Pr(T_{jk} = t_i \mid T_{jk} \geq t_i)$$

and

$$\Pr(T_{jk} = t_i \mid T_{jk} \geq t_i, T_{j,3-k} = t_i) = \theta_i \lambda_i$$

where the constant θ_i is a measure of the dependence between two ears [k and $(3-k)$] of the same child in this risk set R_{2i}. (This same model can also be expressed as a discrete joint distribution; see Le, 1988a.) As shown by Rosner (1982):

$$\hat{\lambda}_i = b_{11i}/b_{1+i}$$

and

$$\text{Var}(\hat{\lambda}_i) = [\lambda_i(1 - \lambda_i)/b_{1+i}](2/e_i)$$

where

$$e_i = 2\lambda_i(1-\lambda_i)/\{\lambda_i(1-\lambda_i) + (\theta_i - 1)\lambda_i^2\}$$

is the "effective number of ears per child" ($1 \le e_i \le 2$). The maximum-likelihood estimators of λ_i and θ_i under $\mathcal{H}_0$ are

$$\hat{\lambda}_{i0} = b_{+1i}/b_{++i}$$

and

$$\hat{\theta}_{i0} = 2b_{++i}c_{+2i}/b_{+1i}^2$$

respectively. These results lead to

$$2/e_i \hat{=} \{2b_{++i}c_{+2i} + b_{+1i}(b_{+0i} - b_{+1i})\}/b_{+1i}b_{+0i}$$

and

$$\widehat{\mathrm{Var}}_0(\hat{\lambda}_i) = b_{+1i}b_{+0i}\{2b_{++i}c_{+2i} + b_{+1i}(b_{+0i} - b_{+1i})\}/b_{++i}^2 b_{1+i}$$

We now consider statistic b_{11i} of Table 3.6, and it can be seen that

$$E_0(b_{11i}) = \frac{b_{+1i}b_{1+i}}{b_{++i}}$$

and, from the above result, the variance of b_{11i} is estimated by

$$\widehat{\mathrm{Var}}_0(b_{11i}) = b_{+1i}b_{+0i}b_{1+i}\{2b_{++i}c_{+2i} + b_{+1i}(b_{+0i} - b_{+1i})\}/b_{++i}^2$$

Test Statistic

Consider now the combined statistic

$$\psi_i = \varphi_{1i}a_{11i} + \varphi_{2i}b_{11i}$$

where φ_{1i} and φ_{2i} are relative weights associated with subsets R_{1i} and R_{2i} respectively; for example, we can take $\varphi_{ji} = 1$ or $\varphi_{ji} = n_{ji}/(n_{1i} + n_{2i})$ for $j = 1, 2$ (other choices are possible; e.g., inverses of estimated variances). The test for comparing group 1 (medical) versus group 2 (control) can then be based on the overall statistic

$$\psi = \sum_{i=1}^{k} w_i \psi_i$$

where w_i is a certain choice of weight associated with time t_i. For example, we can take $w_i = 1$ (similar to the log-rank), or $w_i = n_{1i} + n_{2i}$ (similar to the generalized Wilcoxon or Gehan's test) $w_i = n_{1i} + 2n_{2i}$, among other choices. Under $\mathcal{H}_0$, the standardized statistic

$$z = \{\psi - E_0(\psi)\}/\{\widehat{\text{Var}}_0(\psi)\}^{1/2}$$

is referred to percentiles of the standard normal where

$$E_0(\psi) = \sum_{i=1}^{k} w_i E_0(\psi_i)$$

$$= \sum_{i=1}^{k} w_i [\phi_{1i} E_0(a_{11i}) + \phi_{2i} E_0(b_{11i})]$$

$$\widehat{\text{Var}}_0(\psi) = \sum_{i=1}^{k} w_i^2 \widehat{\text{Var}}_0(\psi_i)$$

$$= \sum_{i=1}^{k} w_i^2 [\varphi_{1i}^2 \widehat{\text{Var}}_0(a_{11i}) + \varphi_{2i}^2 \widehat{\text{Var}}_0(b_{11i})]$$

An application of this proposed method to the data of Appendix C with all weights chosen to be 1, yields $z = 1.69$ (one-sided $p = .05$; and for subsequent comparison, two-sided $p = .09$) while we would have $\chi^2 = 4.05$ (1 df, two-sided $p = .04$) if within-subject correlation is ignored (direct application of the log-rank test). In other words, if within-subject correlation is ignored, the level of significance would be wrongly inflated (calculations were implemented using FORTRAN, supplemented by IMSL subroutines).

As pointed out by Rosner (1982) and Le (1988a), there would be a substantial inflation of the level of significance as the level of dependence between ears increases. To check for this between-ears dependence, the Kendall tau (Weier and Basu, 1980; Oakes, 1982) was calculated as in Section 2.5 resulting in a coefficient of .56 indicating a moderately positive correlation.

The special method in this section is different from two other two-sample tests that have been proposed in the literature because it deals with a different situation. Gail, Santner, and Brown (1980) dealt with "multiple time to tumor" where the recurrences are restricted to the individuals who have experienced the first episode. The method of Wei, Lin, and Weissfeld (1989) addressed the situation involving more than one type of failure, say two types; but there the types are distinguishable, in our situation the two "members" (i.e., ears) are interchangeable.

EXERCISES

3.1. Another way to formulate Gehan's statistic is as follows. Let

$$\theta = \sum_{\text{group 1}} U_i$$

where U_i is the score of the ith subject in sample 1 defined by

$$U_i = L_{1i} - L_{2i}$$

in which

L_{1i} = number of the remaining $(n_1 + n_2 - 1)$ observations that are definitely smaller than the ith

L_{2i} = number of the remaining $(n_1 + n_2 - 1)$ observations that are definitely greater than the ith

For example,

(i) (t_i, δ_i) is *definitely* smaller than (t_j, δ_j) if $\delta_i = 1$ and $t_i \leq t_j$;
(ii) $(t = 3, \delta = 0)$ and $(t = 8, \delta = 1)$ are uncomparable.

Under the null hypothesis, θ is the total of a sample size n, taken from the finite population [size $(n_1 + n_2)$] of the scores U_i's. Find Var(θ) under the null hypothesis of no difference (between samples 1 and 2) and show that this test is reduced to the Wilcoxon test in the absence of censoring.

3.2. Use the following small data set to verify the result of Exercise 3.1, that is, to show that we do get the same standardized statistic from the two different approaches (this scoring procedure and the combination of 2×2 tables using hypergeometric distribution).

$$\text{Sample 1:} \quad 24, 30, 42, 15^+, 40^+, 42^+$$
$$\text{Sample 2:} \quad 10, 26, 28, 30, 41, 12^+$$

3.3. Refer to the data for lung cancer patients in Example 6 of Chapter 1. The two groups—those with no prior therapy and those with prior therapy—were compared in Example 8 of Chapter 1, using the Wilcoxon rank-sum test. Try this comparison again as a combination of 2×2 tables using the hypergeometric distribution. Perform both the log-rank and the generalized Wilcoxon tests; since there are no censored observations, verify that the result of the generalized Wilcoxon test agrees with that of Example 8, Chapter 1.

3.4. Refer to the survival times in Example 13 of Chapter 1. Compare the low-fat and saturated fat diets by both the log-rank and generalized Wilcoxon tests.

3.5. The diet study described in Example 4 of Chapter 1 consists of three groups; survival times are given in Examples 4 and 13 of Chapter 1. Perform a simultaneous comparison of these three diets.

3.6. Refer to the data of Example 4 but ignore the matching design. Compare the two groups and compare the result to that of Example 4.

3.7. Verify that the likelihood ratio (LR) chi-square statistic for the comparison of two exponential distributions is given by

$$\chi^2_{LR} = 2\left[d_1 \ln \frac{d_1}{T_1} + d_2 \ln \frac{d_2}{T_2} - (d_1 + d_2)\ln \frac{d_1 + d_2}{T_1 + T_2}\right]$$

3.8. In Exercise 3.4, we compare the low-fat and saturated-fat diets by performing the two nonparametric tests. We now assume that the survival times are exponentially distributed; try this comparison again by using both the efficient score and likelihood ratio chi-square tests.

EXERCISES

3.9. Suppose we want to compare simultaneously k exponential distributions with parameters $\lambda_1, \lambda_2, \ldots \lambda_k$:

$$\mathcal{H}_0 : \lambda_1 = \lambda_2 = \cdots = \lambda_k$$

Verify that the efficient score and likelihood ratio chi-square statistics are given by

$$\chi^2_{ES} = \left\{ \sum \frac{d_i^2}{T_i} - \frac{(\sum d_i)^2}{\sum T_i} \right\} \left(\sum T_i \bigg/ \sum d_i \right)$$

$$\chi^2_{LR} = 2 \left\{ \sum d_i \ln \frac{d_i}{T_i} - \left(\sum d_i \right) \ln \frac{\sum d_i}{\sum T_i} \right\}$$

respectively; each has $(k - 1)$ degrees of freedom.

3.10. Apply the methods in Exercise 3.9 to the three groups of patients in Example 9 of Chapter 1.

3.11. Try the comparison in Exercise 3.5 again by assuming that survival times are exponentially distributed and by applying the two chi-square tests of Exercise 3.9.

3.12. Refer to the small data set (two samples, each of size 6) in Exercise 3.2. Assuming that the distributions of both samples are exponential, compute both χ^2_{ES} and χ^2_{LR} and compare the result to the log-rank statistic obtained in 3.2.

3.13. Refer back to the clinical trial to evaluate the effect of 6–MP to maintain remission from acute leukemia with (unmatched) data, given in Example 1. Assuming exponential distributions, calculate the "relative risk" of the control group (as compared to 6–MP).

3.14. Refer back to the study of asbestos workers of Example 7 from Chapter 2 (see Table 2.8). Using cancers of the lung as a specific cause of death, we have

Low-to-moderate exposure: $d = 33$
$e = 20.0$
Severe exposure: $d = 87$
$e = 23.2$

(a) Investigate the excess mortality of the severe exposure group (perform the test of significance and calculate a 95 percent confidence interval for the relative risk).

(b) Compare the two levels of exposure.

3.15. Refer back to the small data set of Exercise 3.2. Compare the two samples assuming a piecewise exponential model and compare the results to those obtained by nonparametric methods of Exercise 3.2.

3.16. Refer to the data set on kidney transplants of Appendix A.

(a) Perform these statistical comparisons:
- Males vs. females
- Diabetics vs. nondiabetics
- ALG vs. non-ALG
- With previous transplants vs. without previous transplants

(b) Selecting the subgroup of males who are nondiabetics and have no previous transplants, investigate the effect of ALG. Any positive effects? Short-term/long-term effects?

(c) Divide the data sets into four groups based on age at transplant (< 40, 40–54, 55–64, ≥ 65). Plot the four survival curves and test for a trend effect of age at transplant.

3.17. Refer to the data set on ESRD patients with hemodialysis of Appendix B.

(a) Perform these comparisons:
- Males vs. females
- Nonwhites vs. whites

(b) Divide the data sets into four groups based on duration of dialysis (< 1 year, 1–2 years, 2–3 years, and > 3 years). Plot the four survival curves and test for a trend in survival.

3.18. In a recent study, performed at the Mayo Clinic, of patients having limited Stage II or IIIA ovarian carcinoma, a main goal was to determine whether or not grade of disease was associated with time to progression of disease. At the time of the final analysis of the study, the observations of time to progression of disease (with t^+ denoting a censored observation at time t), for patients with low-grade or well-differentiated cancer, were 28, 89, 175, 195, 309, 377$^+$, 393$^+$, 421$^+$, 447$^+$, 462, 709$^+$, 744$^+$, 770$^+$, 1106$^+$ and 1206$^+$ days. Observation times for patients with high-grade or undifferentiated cancer were 34, 88, 137, 199, 280, 291, 299$^+$, 300$^+$, 309, 351, 358, 369, 369, 370, 375, 382, 392, 429$^+$, 451 and 1119$^+$ days.

EXERCISES

Create your own data file, then write a program to:
(a) Perform the Gehan's and log-rank tests.
(b) Provide the two survival curves in the same graph using the product limit method.
(c) From the results in (b), discuss the power of the tests in (a).

3.19. Consider these two two-piece exponential distributions.

$$\text{Population 1:} \quad \lambda(t) = \begin{cases} \lambda_1 & \text{for } 0 \le t < \pi \\ \lambda_2 & \text{for } \pi \le t \end{cases}$$

$$\text{Population 2:} \quad \lambda(t) = \begin{cases} \rho\lambda_1 & \text{for } 0 \le t < \pi \\ \rho\lambda_2 & \text{for } \pi \le t \end{cases}$$

Given two samples, derive the likelihood function for ρ, λ_1, and λ_2 and the score statistic for testing $\mathcal{H}_0 : gr = 1$. This method can be generalized as follows: The actuarial or life-table method is primarily designed for situations in which actual death and censoring times are not available and only d_i (the number of deaths) and m_i (number of subjects withdrawn) are given for the ith interval. A simple modification for the method is possible when additional information for observed times is known. Suppose that $t_{i1}, t_{i2}, \ldots, t_{in_i}$ $(n_i = d_i + m_i)$ are the observed times in the interval I_i (d_i of these are death times and m_i are censoring times). Suppose the hazard function $\lambda(t)$ is taken to be a step function having the constant value λ_i in the interval I_i; $i = 1, 2, \ldots k$.

(a) For a $t \in I_1$, find $S(t)$; $S(\cdot)$ is the survival function.
(b) Repeat (a) for $t \in I_2$ and $t \in I_i$; $1 \le i \le k$.
(c) Find an estimator $\hat{\lambda}_i$ for λ_i and $\hat{S}(t)$ for $S(t)$; $t \in I_i$.
(d) Suppose we have two large samples of survival data presented in the above layout (life-table set up but individual times are available and stepwise exponential models are assumed). Assume this model for the ith interval:

$$\lambda(t) = \begin{cases} \lambda_{1i} & \text{if group 1} \\ \rho\lambda_{1i} & \text{if group 2} \end{cases}$$

Form the likelihood function for ρ and derive the score test for $\mathcal{H}_0 : \rho = 1$.

$$\lambda(t) = \begin{cases} \lambda_{1i} & \text{if group 1} \\ \rho\lambda_{1i} & \text{if group 2} \end{cases}$$

Form the likelihood function for ρ and derive the score test for $\mathcal{H}_0$: $\rho = 1$.

3.20. Refer to the scoring method of Exercise 3.1. Pool the two samples of sizes n_1, n_2 to form a sample of size $(n_1 + n_2)$, then arrange all the observed times (including censored ones) from smallest to largest. Let n_j be the number of subjects at risk at time t_j, define

$$s_i = \prod_{j=1}^{i} \frac{n_j}{n_j + 1}$$

and assign a score c_i to each subject as follows:

(i) If the subject is the ith event,

$$c_i = 1 - 2s_i$$

(ii) If the subject is a censored observation between the ith and $(i+1)$th events,

$$c_i = 1 - s_i$$

We consider the application of the Siegel–Tukey test to these two sets of scores. What would be the target alternative hypothesis? How is this different from the test for detection of crossing-curves alternatives? Apply the method to the small data set in Exercise 3.2.

CHAPTER 4

Correlation and Regression Analyses

4.1. Simple Regression and Correlation
   Model and Approach
   Measures of Association
   Effects of Measurement Scale
   Tests of Association
4.2. Multiple Regression
   Proportional Hazards Models with Several Covariates
   Testing Hypotheses in Multiple Regression
   Stepwise Procedure
   Estimation of the Baseline Survival Function
4.3. Time-Dependent Covariates
   Examples
   Implementation
   Simple Test of Goodness of Fit
   More on Tests of Goodness of Fit
4.4. Stratification and Its Applications
   Basic Ideas
   Model and Implementation
   Analysis of Epidemiologic Matched Studies
   Homogeneity of Relative Risk in Matched Studies
4.5. A Nontime Application: Evaluation of Confounding Effects in ROC Studies
   Estimator for the ROC Function
   Use of Proportional Hazards Model
   Special Case
Exercises

Some of the most intriguing scientific questions deal with the relationship between two variables, a dependent or response varable Y and an independent variable or covariate X. For example, when both X and Y are continuous measurements, the most common method used to describe their relationship is the simple linear correlation and regression analysis (in which Y is assumed to be normally distributed). When the response variable is a patient's survival time T, represented by observed data $y = (t, \delta)$ with possible censoring—t is the duration time and δ is the survival status indicator—the analysis is even more important; the results are used for identifying prognostic or risk factors and prognosis, that is, the prediction of the future of an individual patient. Such analyses play an important role in medical practice as well as public health intervention.

Example 1. In Example 5 of Chapter 1, data were given for two groups of patients who died of acute myelogenous leukemia. Pa-

(AG Positive) $N = 17$ White Blood Count (WBC)	Survival Time (weeks)	(AG Negative) $N = 16$ White Blood Count (WBC)	Survival Time (weeks)
2,300	65	4,400	56
750	156	3,000	65
4,300	100	4,000	17
2,600	134	1,500	7
6,000	16	9,000	16
10,500	108	5,300	22
10,000	121	10,000	3
17,000	4	19,000	4
5,400	39	27,000	2
7,000	143	28,000	3
9,400	56	31,000	8
32,000	26	26,000	4
35,000	22	21,000	3
100,000	1	79,000	30
100,000	1	100,000	4
52,000	5	100,000	43
100,000	65		

tients were classified into the two groups according to the presence or absence of a morphologic characteristic of white cells. Patients termed AG positive were identified by the presence of Auer rods and/or significant granulature of the leukemic cells in the bone marrow at diagnosis. For the AG-negative patients these factors were absent. Leukemia is a cancer characterized by an overproliferation of white blood cells; the higher the white blood count (WBC), the more severe the disease. Data in the following table clearly suggest that there is such a relationship, and thus when predicting a leukemia patient's survival time, it is realistic to make a prediction dependent on WBC (and any other covariates that are indicators of the progression of the disease).

4.1. SIMPLE REGRESSION AND CORRELATION

In this section we will discuss the basic ideas of simple regression analysis when only one predictor or independent variable is available for predicting the survival of interest; parts of this have been briefly introduced in Section 1.4 and Section 3.1.

Model and Approach

The association between two random variables X and T, the second of which—the survival time T—may be only partially observable due to right censoring, has been the focus of many investigations starting with the historical breakthrough by Cox (1972). The so-called Cox regression model or proportional hazards model (PHM) expresses a log-linear relationship between X and the hazard function of T:

$$\lambda(t \mid X = x) = \lim_{\delta \downarrow 0} \frac{\Pr[t \leq T \leq t + \delta \mid t \leq T, X = x]}{\delta}$$
$$= \lambda_0(t) e^{\beta x}$$

as introduced in Section 1.4 and in Section 3.1. In this model, $\lambda_0(t)$ is an *unspecified* baseline hazard, that is, hazard at $X = 0$, and β is an *unknown* regression coefficient. The estimation of β and subsequent analyses are performed as follows. Denote the ordered distinct death times by

$$t_1 < t_2 < \cdots < t_m$$

and let R_i be the risk set just before time t_i, n_i the number of subjects in R_i, D_i the death set at time t_i, d_i the number of subjects (i.e., deaths) in D_i, and C_i the collection of all possible combinations of subjects from R_i; each combination—or subset of R_i—has d_i members, and D_i is itself one of these combinations. For example, if three subjects (A, B, and C) are at risk just before time t_i and two of them (A and B) die at t_i, then

$$R_i = \{A, B, C\} \quad n_i = 3$$
$$D_i = \{A, B\} \quad d_i = 2$$
$$C_i = \{\{A, B\} = D_i, \{A, C\}, \{B, C\}\}$$

The number of elements in C_i is

$$\binom{n_i}{d_i}$$

Cox (1972) suggests using

$$L = \prod_{i=1}^{m} \Pr(d_i \mid R_i, d_i)$$

$$L = \prod_{i=1}^{m} \frac{\exp(\beta s_i)}{\sum_{C_i} \exp(\beta s_u)}$$

as a likelihood function, called the *partial-likelihood* function in which

$$s_i = \sum_{D_i} x_j$$

$$s_u = \sum_{D_u} x_j \quad D_u \in C_i$$

SIMPLE REGRESSION AND CORRELATION

Cox writes (pp. 190–191; also see Cox, 1975): "Suppose then that $\lambda_0(t)$ is arbitrary. No information can be contributed about β by time intervals in which no failures occur because the components $\lambda_0(t)$ might conceivably be identically zero in such intervals. We therefore argue conditionally on the set of instants at which failures occur; in discrete time we shall condition also on the observed multiplicities. Once we require a method of analysis holding for all $\lambda_0(t)$, consideration of this conditional distribution seems inevitable." With this argument, Cox suggests that we treat his partial likelihood as an ordinary likelihood function; in particular, to find the maximum-likelihood estimate, use the score statistic and the sample information matrix. Of course, the estimation procedure require an iterative method, such as Newton–Raphson. Kalbfleisch and Prentice (1973) justify the use of the partial-likelihood function under the assumption of no ties and Breslow (1974 and 1975) tries with the piecewise exponential baseline. Tsiatis (1981) gives a proof of the asymptotic normality of $\hat{\beta}$ (also see Efron, 1977).

An alternative likelihood, proposed by Peto and Peto (1972), is

$$L = \prod_{i=1}^{m} \frac{\exp(\beta s_j)}{[\sum_{R_i} \exp(\beta x_u)]^{d_i}}$$

which seems to work reasonably well when the number of ties is not excessive and, therefore, becomes rather popular.

Measures of Association

Regression analysis serves two major purposes:

1. Control or intervention
2. Prediction

In many studies, such as the one in Example 1, one important objective is establishing a statistical relationship between survival time and independent variables or covariates measured from patients; findings may lead to important decisions in patient management and/or public health interventions. Of course, if such a relationship is established, then it is more realistic to make prediction concerning a patient's survival probabilities conditional on his/her covariate values, for example, WBC. Effects of a factor on a patient's survival are measured by the relative risk.

We first consider the case of a binary covariate with the conventional coding

$$X_i = \begin{cases} 0 & \text{if the patient is not exposed} \\ 1 & \text{if the patient is exposed} \end{cases}$$

Here, the term "exposed" may refer to a risk factor such as smoking, or a patient's characteristic such as race (white/nonwhite) or sex (male/female). It can be seen that, from the proportional hazards model,

$$\lambda(t; \text{nonexposed}) = \lambda_0(t)$$

$$\lambda(t; \text{exposed}) = \lambda_0(t)e^\beta$$

So that the ratio

$$e^\beta = \frac{\lambda(t; \text{exposed})}{\lambda(t; \text{nonexposed})}$$

represents the relative risk (RR) of the exposure, exposed vs. nonexposed. In other words, the regression coefficient β is the value of the relative risk on the log scale

Similarly, we have for a continuous covariate X and any value x of X,

$$\lambda(t; X = x) = \lambda_0(t)e^{\beta x}$$

$$\lambda(t; X = x + 1) = \lambda_0(t)e^{\beta(x+1)}$$

So that the ratio

$$e^\beta = \frac{\lambda(t; X = x + 1)}{\lambda(t; X = x)}$$

represents the relative risk due to *one unit increase* in the value of X, $X = x + 1$ vs. $X = x$. For example, a systolic blood pressure of 114 mmHg vs. 113 mmHg. For m units increase in the value of X, say $X = x + m$ vs. $X = x$, the corresponding relative risk is $e^{m\beta}$.

SIMPLE REGRESSION AND CORRELATION

The regression coefficient β can be estimated iteratively using the first and second derivatives of the partial-likelihood function. From the results, we can obtain a point estimate

$$\boxed{\widehat{RR} = e^{\hat{\beta}}}$$

and its 95 percent confidence interval

$$\exp[\hat{\beta} \pm 1.96 \operatorname{SE}(\hat{\beta})]$$

Effects of Measurement Scale

It should be noted that the relative risk, used as a measure of association between survival time and a covariate, depends on the coding scheme for a binary factor and, for a continuous covariate X, the scale with which to measure X. For example, if we use the following coding for a factor

$$X_i = \begin{cases} -1 & \text{if the patient is not exposed} \\ 1 & \text{if the patient is exposed} \end{cases}$$

then

$$\lambda(t; \text{nonexposed}) = \lambda_0(t)e^{-\beta}$$
$$\lambda(t; \text{exposed}) = \lambda_0(t)e^{\beta}$$

So that

$$RR = \frac{\lambda(t; \text{exposed})}{\lambda(t; \text{nonexposed})}$$
$$= e^{2\beta}$$

and its 95 percent confidence interval

$$\exp[2(\hat{\beta} \pm 1.96 \operatorname{SE}(\hat{\beta}))]$$

Of course, the estimate of β under the new coding scheme is only half of that under the former scheme; therefore, the estimate of the RR remains unchanged.

The following example, however, will show the clear effect of measurement scale in the case of a continuous measurement.

Example 2. Refer to the data for patients with acute myelogenous leukemia in Example 1 and suppose we want to investigate the relationship between survival time of AG-positive patients and WBC in two different ways using either (i) $X = $ WBC or (ii) $X = \log(\text{WBC})$.

1. For $X = $ WBC, we find

$$\hat{\beta} = .0000167$$

from which, the relative risk for (WBC = 100,000) vs. (WBC = 50,000) would be

$$\text{RR} = \exp[(100{,}000 - 50{,}000)(.0000167)]$$
$$= 2.31$$

2. For $X = \log(\text{WBC})$, we find

$$\hat{\beta} = .612331$$

from which, the relative risk for (WBC = 100,000) vs. (WBC = 50,000) would be

$$\text{RR} = \exp\{[\log(100{,}000) - \log(50{,}000)][.612331]\}$$
$$= 1.53$$

Note: An SAS program would include these instructions:

```
PROC PHREG DATA=CANCER;
MODEL WEEKS*DEATH(0)=WBC;
```

where CANCER is the name assigned to the data set, WEEKS is the variable name for duration time, DEATH the variable name for survival status, and "0" is the coding for censoring.

The above results are different for two different choices of X, and this causes an obvious problem of choosing an appropriate measurement scale. Of course, we assume a *linear* model and one choice of X would fit better than the other.

Tests of Association

The last two sections dealt with inferences concerning the regression coefficient β, including both point and interval estimation of this parameter and the relative risk. In this section, we take up the test of significance approach to regression analysis. The null hypothesis to be considered is:

$$\mathcal{H}_0 : \beta = 0$$

The reason for interest in testing whether or not $\beta = 0$ is that $\beta = 0$ implies there is no relation between survival time T and the covariate X under investigation.

Recall that under the proportional hazards model

$$\lambda(t) = \lim_{\delta \downarrow 0} \frac{\Pr[t \leq T \leq t + \delta \mid t \leq T]}{\delta}$$

$$= \lambda_0(t) e^{\beta x}$$

the partial-likelihood function is

$$L = \prod_{i=1}^{m} \Pr(D_i \mid R_i, d_i)$$

$$= \prod_{i=1}^{m} \frac{\exp(\beta s_i)}{\sum_{c_i} \exp(\beta s_u)}$$

in which

$$s_i = \sum_{D_i} x_j$$

$$s_u = \sum_{D_u} x_j \qquad D_u \in C_i$$

where R_i is the risk set just before time t_i, n_i the number of subjects in R_i, D_i the death set at time t_i, d_i the number of subjects (i.e., deaths) in D_i, and C_i the collection of all possible combinations of subjects from R_i; each combination—or subset of R_i—has d_i members, and D_i is itself one of these combinations. By treating the above as a regular likelihood function, we have

$$\frac{d}{d\beta} \log L = \sum_{i=1}^{m} \left\{ s_i - \frac{\sum s_u \exp(\beta s_u)}{\sum_{C_i} \exp(\beta s_u)} \right\}$$

$$\stackrel{\mathcal{H}_0}{=} \sum_{i=1}^{m} \{s_i - \mu_i\} \qquad \text{setting} \quad \beta = 0$$

where μ_i is the mean of s over C_i:

$$\mu_i = \sum_{C_i} s_u / N_i$$

and

$$-\frac{d^2}{d\beta^2} \log L = \sum_{i=1}^{m} \left\{ \frac{\sum s_u^2 \exp(\beta s_u)}{\sum_{C_i} \exp(\beta s_u)} - \left[\frac{\sum s_u \exp(\beta s_u)}{\sum_{C_i} \exp(\beta s_u)} \right]^2 \right\}$$

$$\stackrel{\mathcal{H}_0}{=} \sum_{i=1}^{m} \left\{ \sum_{C_i} s_u^2 / N_i - \mu_i^2 \right\} \qquad \text{setting} \quad \beta = 0$$

$$= \sum_{i=1}^{m} \sigma_i^2$$

SIMPLE REGRESSION AND CORRELATION   **171**

where σ_i^2 is the variance of s over C_i (considered as a finite population). This procedure leads to a score statistic:

$$z = \left\{\sum_{i=1}^{m}(s_i - \mu_i)\right\} \bigg/ \left\{\sum_{i=1}^{m}\sigma_i^2\right\}^{1/2}$$

In this formulation of the score statistic for testing the null hypothesis:

$$\mathcal{H}_0 : \beta = 0$$

we made no assumptions on the covariate X; different measurements scales will be now considered.

Categorical Factors

We first consider the case of a binary covariate with the conventional coding

$$x_i = \begin{cases} 0 & \text{if the patient is not exposed} \\ 1 & \text{if the patient is exposed} \end{cases}$$

It is obvious that, for each element in C_i (the collection of all possible combinations of subjects from R_i having d_i members), s is the value of a random variable counting the number of exposed patients. Under $\mathcal{H}_0$, s is distributed as *hypergeometric*. Therefore, the test based on the above score statistic is identical to the log-rank test comparing two groups, those patients who were exposed vs. those patients who were not exposed. Similarly, the case of a nominal factor leads to the log-rank test for the comparison of several groups, or the Cox–Mantel test of Chapter 3.

Consider now the case of an ordinal covariate with R levels. If we assign to all subjects in level i, $1 \leq i \leq R$, the same score x_i and apply the above test, then it leads to the Tarone test for survival trends as introduced and discussed in the last section of Chapter 3.

Continuous Factors

As for continuous factors, the same log-rank test applies as well (of course without the formulation of those 2×2 tables).

TABLE 4.1. Some Hypothetical Data

Subject	i	t_i	Status	Covariate x
a	1	1	D	6
b	2	2	D	9
c		8	A	12
d	3	10	D	8
e	3	10	D	10
f	4	14	D	13
g		15	A	14

Example 3. To illustrate the procedure, consider the data in Table 4.1. We have, for example at $t = 10$, the following s values:

$$8 + 10 = 18, \quad 8 + 13 = 21, \quad 8 + 14 = 22,$$
$$10 + 13 = 23, \quad 10 + 14 = 24, \quad 13 + 14 = 27$$

leading to:

$$\mu_i = (18 + 21 + 22 + 23 + 24 + 27)/6$$
$$= 22.5$$
$$\sigma_i^2 = (18^2 + 21^2 + 22^2 + 23^2 + 24^2 + 27^2)/6 - (22.5)^2$$
$$= 7.58$$

The rest of the log-rank analysis is summarized in Table 4.2. From these figures, an application of the log-rank test yields

$$z = \frac{(6 - 10.29) + (9 - 11.0) + (18 - 22.5) + (13 - 13.5)}{\{7.06 + 4.67 + 7.58 + .25\}^{1/2}}$$
$$= -2.574$$

indicating a significant association.

TABLE 4.2. Log-Rank Analysis for Data of Table 4.1

t_i	x_i	s_i	Mean of x over R_i	Variance of x over R_i	Mean of s over C_i	Variance of s over C_i
1	6		10.29	7.06		
2	9		11.0	4.67		
10	8, 10	18			22.5	7.58
14	13		13.5	.25		

Note: For a sample SAS program, include these instructions:

```
PROC LIFETEST METHOD=KM;
TIME WEEKS*FAILURE(0);
TEST COVAR;
```

where KM stands for Kaplan–Meier method, WEEKS is the variable name for duration time, FAILURE the variable name for survival status, "0" is the coding for censoring, and COVAR is the variable name for the covariate.

Nonparametric Tests

The proportional hazards model has become the model of choice for many studies leading to the above log-rank test in view of inference that is rank-invariant with respect to survival times (in addition to the flexibility of an unspecified baseline hazard). However, since the invariance property does not apply to covariate values, the log-rank test is only a semi-nonparametric procedure the result of which depends on the scale with which to measure the covariate X (Le and Grambsch, 1994; O'Quigley and Prentice, 1991).

Example 4. Refer to the data for patients with acute myelogenous leukemia in Example 1 and suppose we want to apply the log-rank procedure to investigate the relationship between survival time of AG-positive patients and WBC in two different ways using either (i) $X = $ WBC or (ii) $X = \log(\text{WBC})$.

1. For $X = $ WBC, we find

$$z = 2.56$$

2. For $X = \log(WBC)$, we find

$$z = 3.01$$

Both results indicate a highly significant relationship; however, the levels of significant are different.

A close examination of the z statistic in the log-rank procedure readily leads to a nonparametric procedure. We rank the s values obtained from members of C_i from 1 to N_i; under random assignment, that is, no covariate effects, we would have the mean of ranks

$$\mu_i = (N_i + 1)/2$$

and variance

$$\sigma_i^2 = (N_i^2 - 1)/12$$

Let r_i be the rank of s_i (obtained from death set D_i; $1 \leq r_i \leq N_i$), a nonparametric test can be formed based on the following statistic:

$$z = \left\{ \sum_{i=1}^{k} [r_i - (N_i + 1)/2] \right\} \bigg/ \left\{ \sum_{i=1}^{k} [(N_i^2 - 1)/12] \right\}^{1/2}$$

Example 5. Using the hypothetical data of Table 4.1, a nonparametric analysis is summarized in Table 4.3; for example, at

TABLE 4.3. Nonparametric Analysis for Data of Table 4.1

t_i	Rank of x_i	Rank of s_i	Ave. Rank of x over R_i	Variance over R_i	Ave. Rank of s over C_i	Variance over C_i
1	1		4	4.0		
2	2		3.5	2.92		
10		1			3.5	2.92
14	1		1.5	.25		

$t = 10$, the rank of $s_i = 18$ is 1 whereas

$$\mu_i = (6+1)/2$$
$$= 3.5$$
$$\sigma_i^2 = (36-1)/12$$
$$= 2.92$$

From these figures, an application of the nonparametric test yields

$$z = \frac{(1-4) + (2-3.5) + (1-3.5) + (1-1.5)}{\{4.0 + 2.92 + 2.92 + .25\}^{1/2}}$$
$$= -2.361$$

a result which is similar to that of the log-rank test in the previous example.

Note: No packaged programs are available for this test.

An alternative approach to the above rank test is first applying the Wilcoxon test to subjects in the risk set R_i (deaths vs. survivors) and then pool the results over times. In other words, we rank the x *values* obtained from members of R_i from 1 to n_i. Let r_i be the *sum of the ranks* obtained from members of D_i (which is treated, under $\mathcal{H}_0$, as a sample of size d_i). Under random assignment that is, no covariate effects, we have the mean

$$\mu_i = d_i(n_i + 1)/2$$

and variance

$$\sigma_i^2 = d_i(n_i - d_i)(n_i + 1)/12$$

The decision is then based on the z *statistic*:

$$z = \left\{ \sum_{i=1}^{k} [r_i - d_i(n_i + 1)/2] \right\} \bigg/ \left\{ \sum_{i=1}^{k} [d_i(n_i - d_i)(n_i + 1)/12] \right\}^{1/2}$$

In the absence of ties, the two nonparametric tests are identical. If there are tied observations, the former test is based on the *rank of the sum* and the latter version is based on *the sum of the ranks*; results are close but need not be identical.

Example 6. Using the hypothetical data of Table 4.1, we have at $t = 10$, the sum of the ranks is $r_i = 1 + 2 = 3$ whereas

$$\mu_i = (2)(4+1)/2$$

$$= 5.0$$

$$\sigma_i^2 = (2)(2)(4+1)/12$$

$$= 1.67$$

From these and other figures, an application of the new test yields

$$z = \frac{(1-4)+(2-3.5)+(3-5)+(1-1.5)}{\{4.0+2.92+1.67+.25\}^{1/2}}$$

$$= -2.354$$

An additional advantage of nonparametric procedures is that they can be modified to investigate nonmonotonic associations; for example, we can use them in conjunction with the Siegel–Tukey ranking of Chapter 1.

4.2. MULTIPLE REGRESSION

The effect of some factor on survival time may be influenced by the presence of other factors through effect modifications, that is, interactions. Therefore, in order to provide a more comprehensive prediction of the future of patients with respect to duration, course, and outcome of a disease, it is very desirable to consider a large number of factors and sort out which ones are most closely related to diagnosis. In this section, we will discuss a multivariate method for risk determination. This method, which is multiple regression analysis, involves a linear combination of the explanatory or independent variables; the variables must be quantitative with particular numerical values for each patient. Information concerning possible factors are usually obtained as a subsidiary aspect from clinical trials that were designed to compare treatments. A covariate or prognostic patient characteristic may be dichotomous, polytomous, or continuous (categorical factors will be represented by dummy variables). Examples of dichotomous covariates are sex, and presence or absence of certain comorbidity. Polytomous covariates include race and different grades of symptoms; these can be covered by the use of *dummy variables*. Continous covariates include patient age, blood pressure, and the like. In many cases, data transformations (e.g., taking the logarithm) may be desirable.

Proportional Hazards Models with Several Covariates

Suppose we want to consider k covariates simultaneously, the proportional hazards model of the previous section can be easily generalized and expressed as:

$$\lambda[t \mid \mathbf{X} = (x_1, x_2, \ldots, x_k)] = \lim_{\delta \downarrow 0} \frac{\Pr[t \leq T \leq t + \delta \mid t \leq T, \mathbf{X} = \mathbf{x}]}{\delta}$$

$$= \lambda_0(t) e^{[\beta^T \mathbf{x}]}$$

$$= \lambda_0(t) e^{[\beta_1 x_1 + \beta_2 x_2 + \cdots + \beta_k x_k]}$$

where $\lambda_0(t)$ is an *unspecified* baseline hazard, that is, hazard at $\mathbf{X} = 0$, and $\beta^T = (\beta_1, \beta_2, \ldots, \beta_k)$ are k *unknown* regression coefficients. In order to have a meaningful baseline hazard, it may be necessary to standardize continuous covariates about their means:

$$\text{New } x = x - \bar{x}$$

so that $\lambda_0(t)$ is the hazard function associated with *a typical patient* (i.e., a hypothetical one having all covariates at their average values).

The estimation of β and subsequent analyses are performed similar to the univariate case using the Cox partial-likelihood function:

$$L = \prod_{i=1}^{m} \Pr(d_i \mid R_i, d_i)$$

$$= \prod_{i=1}^{m} \frac{\exp[\sum_{j=1}^{k} \beta_j s_{ji}]}{\sum_{C_i} \exp[\sum_{j=1}^{k} \beta_j s_{ju}]}$$

where

$$s_{ji} = \sum_{l \in D_i} x_{jl}$$

$$s_{ju} = \sum_{l \in D_u} x_{jl} \quad D_u \in C_i$$

Also similar to the univariate case, $\exp(\beta_i)$ represents:

(i) The relative risk associated with an exposure if X_i is binary (exposed $X_i = 1$ vs. unexposed $X_i = 0$) or
(ii) The relative risk due to one unit increase if X_i is continuous ($X_i = x + 1$ vs. $X_i = x$)

and after $\hat{\beta}_i$ and its standard error have been obtained, a 95 percent confidence interval for the above relative risk is given by:

$$\exp[\hat{\beta}_i \pm 1.96 \text{SE}(\hat{\beta}_i)]$$

These results are necessary in the effort to identify important prognostic or risk factors. Of course, before such analyses are done, the problem and the data have to be examined carefully. If some of the variables are highly correlated, then one or fewer of the correlated factors are likely to be as good predictors as all of them; information from other similar studies also has to be incorporated so as to drop some of these correlated explanatory variables. The uses of products, such as $X_1 X_2$, and higher power terms, such as X_1^2, may be necessary and can improve the goodness of fit. It is important to note that we are assuming a *linear* regression model in which, for example, the relative risk due to one unit increase in the value of a continuous X_i ($X_i = x + 1$ vs. $X_i = x$) is independent of x. Therefore, if this *linearity* seems to be violated, the incorporation of powers of X_i should be seriously considered. The use of products will help in the investigation of possible effect modifications. And, finally, the messy problem of missing data; most packaged programs would delete the patient if one or more covariate values are missing.

Example 7. Appendix A gives the data for 469 patients with kidney transplants. The primary endpoint was graft survival and time to graft failure was recorded in months. This study included measurements of many covariates that may be related to survival experience. Eight such variables are given in this data set: age (years) at transplant, sex, duration (months) of dialysis prior to transplant, diabetes, prior transplant (yes/no), blood transfusion (blood units), mismatch score, and use of ALG (an immune suppression drug). An application of the Cox proportional hazards multiple regression is shown in Table 4.4. The three important variables in predicting survival time of the kidney graft were age, use of ALG, and diabetes; the use of ALG could reduce the risk of graft failure by about 46%.

TABLE 4.4. Multiple Regression Analysis of Kidney Transplant Data

Factor	Coefficient	St. Error	Relative Risk	95% C.I.
Age	0.01726	0.00610	1.017	(1.005,1.030)
Sex	−0.02986	0.14882	0.971	(0.725,1.299)
Dialysis	0.00273	0.00468	1.003	(0.994,1.012)
Diabetes	0.34849	0.19237	1.417	(0.972,2.066)
Prior transplants	−0.10922	0.24160	0.897	(0.558,1.440)
Blood transfusion	0.00391	0.00498	1.004	(0.994,1.014)
Mismatch score	0.08277	0.09179	1.086	(0.907,1.300)
ALG	−0.61281	0.17123	0.542	(0.387,0.769)

Note: An SAS program for use with three covariates: Age, Sex, and Diabetes would include these instructions:

```
PROC PHREG DATA=KIDNEY;
MODEL MONTHS*FAILURE(0)=AGE SEX DIABETES;
```

where KIDNEY is the name assigned to the data set, MONTHS is the variable name for duration time, FAILURE the variable name for survival status of the new graft, and "0" is the coding for censoring.

Effect Modifications

Consider the model:

$$\lambda[t \mid \mathbf{X} = (x_1, x_2)] = \lambda_0(t) e^{[\beta_1 x_1 + \beta_2 x_2 + \beta_3 x_1 x_2]}$$

The meaning of β_1 and β_2 here is not the same as that given earlier because of the cross-product term $\beta_3 x_1 x_2$. Suppose that both X_1 and X_2 are binary, then:

1. For $X_2 = 1$ or exposed, we have

$$\lambda(t; \text{ not exposed to } X_1) = \lambda_0(t) e^{\beta_2}$$
$$\lambda(t; \text{ exposed to } X_1) = \lambda_0(t) e^{\beta_1 + \beta_2 + \beta_3}$$

So that the ratio

$$e^{\beta_1 + \beta_3} = \frac{\lambda(t; \text{ exposed})}{\lambda(t; \text{ nonexposed})}$$

represents the relative risk of the exposure to X_1, exposed vs. nonexposed, in the presence of X_2, whereas

2. For $X_2 = 0$ or not exposed, we have

$$\lambda(t; \text{ not exposed to } X_1) = \lambda_0(t)$$
$$\lambda(t; \text{ exposed to } X_1) = \lambda_0(t) e^{\beta_1}$$

So that the ratio

$$e^{\beta_1} = \frac{\lambda(t; \text{ exposed})}{\lambda(t; \text{ nonexposed})}$$

represents the relative risk of the exposure to X_1, exposed vs. nonexposed, in the absence of X_2. In other words, the effect of X_1 depends on the level (presence or absence) of X_2 and vice versa. This phenomenon is called *effect modification*, that is, one factor modifies the effect of the other. The cross-product term $x_1 x_2$ is called an interaction term. The use of these products will help in the investigation of possible effect modifications.

Polynomial Regression

Consider the model:

$$\lambda(t \mid X = x) = \lambda_0(t) e^{[\beta_1 x + \beta_2 x^2]}$$

where X is a continuous covariate. The meaning of β_1 here is not the same as that given earlier because of the quadratic term $\beta_2 x^2$. We have, for example,

$$\lambda(t; X = x) = \lambda_0(t) e^{\beta_1 x + \beta_2 x^2}$$

$$\lambda(t; X = x + 1) = \lambda_0(t) e^{\beta_1 (x+1) + \beta_2 (x+1)^2}$$

So that the relative risk

$$\begin{aligned} \text{RR} &= \frac{\lambda(t; X = x + 1)}{\lambda(t; X = x)} \\ &= \exp[\beta_1 + \beta_2(2x + 1)] \\ &= \text{RR}(x) \end{aligned}$$

is function of x.

Polynomial models with an independent variable present in higher powers than the second are not often used. The second-order or quadratic model has two basic type of uses: (i) when the true relationship is a second-degree polynomial or when the true relationship is unknown but the second-degree polynomial provides a better fit than a linear one, but (ii) more often, a quadratic model is fitted for the purpose of establishing the linearity. The key item to look is whether $\beta_2 = 0$.

The use of polynomial models is not without drawbacks. The most potential drawback is that multicolinearity is unavoidable; especially, if the covariate is restricted to a narrow range, then the degree of multicolinearity can be quite high. Another problem arises when one wants to use the stepwise regression search method. In addition, finding a satisfactory interpretation for the *curvature* effect coefficient β_2 is not easy.

Testing Hypotheses in Multiple Regression

Once we have fit a multiple proportional hazards regression model and obtained estimates for the various parameters of interest, we want to answer questions about the contributions of various factors to the prediction of the future of patients. There are three types of such questions:

(i) An overall test: Taken collectively, does the entire set of explanatory or independent variables contribute significantly to the prediction of survivorship?
(ii) Test for the value of a single factor: Does the addition of one particular factor of interest add significantly to the prediction of survivorship over and above that achieved by other factors?
(iii) Test for contribution of a group of variables: Does the addition of a group of factors add significantly to the prediction of survivorship over and above that achieved by other factors?

Overall Regression Tests

We now consider the first question stated above concerning an overall test for a model containing k factors, say,

$$\lambda[t \mid \mathbf{X} = (x_1, x_2, \ldots, x_k)] = \lim_{\delta \downarrow 0} \frac{\Pr[t \leq T \leq t + \delta \mid t \leq T, \mathbf{X} = \mathbf{x}]}{\delta}$$

$$= \lambda_0(t) e^{[\beta^T \mathbf{x}]}$$

$$= \lambda_0(t) e^{[\beta_1 x_1 + \beta_2 x_2 + \cdots + \beta_k x_k]}$$

MULTIPLE REGRESSION

The null hypothesis for this test may be stated as: "all k-independent variables *considered together* do not explain the variation in survival times." In other words,

$$\mathcal{H}_0 : \beta_1 = \beta_2 = \cdots = \beta_k = 0$$

Three likelihood-based statistics can be used to test this *global* null hypothesis; each has an symptotic chi-squared distribution with k degrees of freedom under $\mathcal{H}_0$.

(i) Likelihood test:

$$\chi^2_{LR} = 2[\ln L(\hat{\beta}) - \ln L(0)]$$

(ii) Wald test:

$$\chi^2_W = \hat{\beta}^T [\hat{V}(\hat{\beta})]^{-1} \hat{\beta}$$

(iii) Score test:

$$\chi^2_S = \left[\frac{\delta \ln L(0)}{\delta \beta}\right] \left[-\frac{\delta^2 \ln L(0)}{\delta \beta^2}\right]^{-1} \left[\frac{\delta \ln L(0)}{\delta \beta}\right]$$

All three statistics are provided by most standard computer programs.

Example 8. Refer to the data set on kidney transplants of Example 7 with eight covariates. We have the following test statistics for the global null hypothesis:

(i) Likelihood test:

$$\chi^2_{LR} = 23.055 \quad \text{with} \quad 8 \text{ dfs} \quad p = .0033$$

(ii) Wald test:

$$\chi^2_W = 23.709 \quad \text{with} \quad 8 \text{ dfs} \quad p = .0026$$

(iii) Score test:

$$\chi^2_S = 24.108 \quad \text{with} \quad 8 \text{ dfs} \quad p = .0022$$

Tests for a Single Variable

Let us assume that we now wish to test whether the addition of one particular factor of interest adds significantly to the prediction of survivorship over and above that achieved by other factors already present in the model. The null hypothesis for this test may be stated as: "Factor X_i does not have any value added to the prediction of survivorship *given that other factors are already included in the model.*" In other words,

$$\mathcal{H}_0 : \beta_i = 0$$

To test such a null hypothesis, one can perform a likelihood ratio chi-squared test, with 1 df, similar to that for the above global hypothesis:

$$\chi^2_{LR} = 2[\ln L(\hat{\beta}; \text{all } X\text{'s}) - \ln L(\hat{\beta}; \text{all other } X\text{'s with } X_i \text{ deleted})]$$

A much easier alternative method is using:

$$\boxed{z_i = \frac{\hat{\beta}_i}{\text{SE}(\hat{\beta}_i)}}$$

where $\hat{\beta}_i$ is the corresponding estimated regression coefficient and $\text{SE}(\hat{\beta}_i)$ is the estimate of the standard error of $\hat{\beta}_i$, both of which are printed by standard packaged programs. In performing this test, we refer the value of the z statistic to percentiles of the standard normal distribution. This is equivalent to the Wald chi-squared test as applied to one parameter.

TABLE 4.5. Multiple Regression Analysis of Kidney Transplant Data

Factor	Coefficient	St. Error	z Statistic	p Value
Age	0.01726	0.00610	2.831	0.0046
Sex	−0.02986	0.14882	−0.201	0.8410
Dialysis	0.00273	0.00468	0.584	0.5594
Diabetes	0.34849	0.19237	1.812	0.0700
Prior transplant	−0.10922	0.24160	−0.452	0.6512
Blood transfusion	0.00391	0.00498	0.785	0.4326
Mismatch score	0.08277	0.09179	0.902	0.3672
ALG	−0.61281	0.17123	−3.579	0.0003

Example 9. Refer to the data set on kidney transplants of Example 7 with eight covariates, shown in Table 4.5. The effects of age and use of ALG are significant at the 5 percent level whereas the effect of diabetes is marginally significant ($p = .07$).

Example 10. Refer again to the data set on kidney transplants of Appendix A and Example 7 but this time we investigate only one covariate, the age of patients. After fitting the second-degree polynomial model,

$$\lambda(t \mid \text{Age}) = \lambda_0(t) e^{[\beta_1(\text{Age}) + \beta_2(\text{Age})^2]}$$

we obtained the results indicating that the *curvature effect* is not statistically significant ($p = .4573$) (Table 4.6). However, the result should be interpreted cautiously; there were few older patients and the insignificance may be a result of strong multicolinearity.

TABLE 4.6. Polynomial Regression Analysis

Factor	Coefficient	St. Error	z Statistic	p Value
Age	0.04273	0.03597	1.118	0.2348
Age^2	−0.00032	0.00043	−0.743	0.4573

Example 11. Refer to the myelogenous leukemia data of Example 1. Patients were classified into the two groups according to

TABLE 4.7. Investigation of Effect Modification

Factor	Coefficient	St. Error	z Statistic	p Value
WBC	0.14654	0.17869	0.821	0.4122
AG group	−5.85637	2.75029	−2.129	0.0332
Product	0.49527	0.27648	1.791	0.0732

the presence or absence of a morphologic characteristic of white cells and the primary covariate is WBC. Using

$$X_1 = \ln(\text{WBC})$$

$$X_2 = \text{AG-group}; \quad 0 \text{ if negative and } 1 \text{ if positive}$$

we fit the following model with one interaction term:

$$\lambda[t \mid \mathbf{X} = (x_1, x_2)] = \lambda_0(t) e^{[\beta_1 x_1 + \beta_2 x_2 + \beta_3 x_1 x_2]}$$

From the results given in Table 4.7 it can be seen that the interaction effect is almost significant at the 5 percent level ($p = .0732$); that is, the presence of the morphologic characteristic modifies substantially the effect of WBC.

Contribution of a Group of Variables

This testing procedure addresses the more general problem of assessing the additional contribution of two or more factors to the prediction of survivorship over and above that made by other variables already in the regression model. In other words, the null hypothesis is of the form

$$\mathcal{H}_0 : \beta_1 = \beta_2 = \cdots = \beta_m = 0$$

To test such a null hypothesis, one can perform a likelihood ratio chi-squared test, with m df,

$$\chi^2_{LR} = 2[\ln L(\hat{\beta}; \text{ all } X\text{'s})$$
$$- \ln L(\hat{\beta}; \text{ all other } X\text{'s with } X\text{'s under investigation deleted})]$$

As with the above *z test*, this *multiple contribution* procedure is very useful for assessing the importance of potential explanatory variables. In particular it is often used to test whether a similar group of variables, such as *demographic characteristics*, is important for the prediction of survivorship; these variables have some trait in common. Another application would be a collection of powers *and/or* product terms (referred to as interaction variables). It is often of interest to assess the interaction effects collectively before trying to consider individual interaction terms in a model. In fact, such use reduces the total number of tests to be performed, and this, in turn, helps to provide better control of overall type I error rates that may be inflated due to multiple testing.

Example 12. Refer to the data set on kidney transplants of Example 7 with eight covariates, and we consider, collectively, these three interaction terms: Diabetes∗Dialysisduration, Diabetes∗Mismatch score, Diabetes∗Use of ALG.

1. With the original 8 variables, $\ln L = -1051.557$
2. With all 11 variables, we obtained: $\ln L = -1051.098$

Therefore:

$$\chi^2_{LR} = 2[\ln L(\hat{\beta};\text{ eleven factors}) - \ln L(\hat{\beta};\text{ eight original factors})]$$

$$= .918; \quad 3 \text{ dfs}, \quad p = .821$$

In other words, all three interaction terms, *considered together*, do not explain the variation in survival times or that diabetes does not alter the effects of the other three factors.

Stepwise Procedure

In applications, our major interest is to identify important prognostic factors. In other words, we wish to identify from many available factors a small subset of factors that relate significantly to the length of survival time of patients. In that identification process, of course, we wish to avoid a type I (false positive) error. In a regression analysis, a type I error corresponds to including a predictor that has no real relationship to survivorship; such an inclusion can greatly confuse the interpretation of the regression results. In a standard multiple regression analysis, this goal can be achieved by using a strategy that adds into or removes from a regression model one factor at a time according to certain order of relative importance. Therefore the two important steps are:

1. Specifying a criterion or criteria for selecting a model.
2. Specifying a strategy for applying the chose criterion or criteria.

Strategies

This is concerned with specifying the strategy for selecting variables. Traditionally, such a strategy is concerned with which and whether a particular variable should be added to a model or whether which or any variable should be deleted from a model at a particular stage of the process. As computers became more accessible and more powerfull, these practices became more popular.

Forward Selection Procedure

In the forward selection procedure, we proceed as follows:

Step 1: Fit a simple linear regression model to each factor, one at a time.
Step 2: Select the most important factor according to certain predetermined criterion.
Step 3: Test for the significance of the factor selected in step 2 and determine, according to certain predetermined criterion, whether or not to add this factor to the model.
Step 4: Repeat steps 2 and 3 for those variables not yet in the model. At any subsequent step, if none meets the criterion in step 3, no more variables are included in the model and the process is terminated.

Backward Elimination Procedure

In the backward elimination procedure, we proceed as follows:

Step 1: Fit the multiple regression model containing all available independent variables.

Step 2: Select the least important factor according to certain predetermined criterion; this is done by considering one factor at a time and treat it as though it were the last variable to enter.

Step 3: Test for the significance of the factor selected in step 2 and determine, according to certain predetermined criterion, whether or not to delete this factor from the model.

Step 4: Repeat steps 2 and 3 for those variables still in the model. At any subsequent step, if none meets the criterion in step 3, no more variables are removed in the model and the process is terminated.

Stepwise Regression Procedure

Stepwise regression is a modified version of forward regression that permits reexamination, at every step, of the variables incorporated in the model in previous steps. A variable entered at an early stage may become superfluous at a later stage because of its relationship with other variables now in the model; the information it provides becomes redundant. That variable may be removed, if meeting the elimination criterion, and the model is re-fitted with the remaining variables, and the forward process goes on. The whole process, one step forward followed by one step backward, continues until no more variables can be added or removed.

Criteria

For the first step of the forward selection procedure, decisions are based on individual score test results (chi-squared, 1 df). In subsequent steps, both forward and backward, the ordering of levels of importance (step 2) and the selection (test in step 3) are based on the likelihood ratio chi-squared statistic:

$$\chi^2_{LR} = 2[\ln L(\hat{\beta}; \text{ all other } X\text{'s}) - \ln L(\hat{\beta}; \text{ all other } X\text{'s with one } X \text{ deleted})]$$

Example 13. Refer to the data set on kidney transplants of Example 7 and Appendix A with eight covariates: age (years) at transplant, sex, duration (months) of dialysis prior to transplant, diabetes, prior transplant (yes/no), blood transfusion (blood units) mismatch score, and use of ALG (an immune suppression drug). This time we perform a stepwise regression analysis in which we specify that a variable has to be significant at the 0.10 level before it can enter into the model and that a variable in the model has to be significant at the 0.15 for it to remain in the model (most standard computer programs allow users to make these selections; default values are available). First, we get these individual score test results for all variables:

Variable	Score χ^2	p Value
Age	8.0530	0.0045
Sex	0.0708	0.7902
Dialysis duration	2.1295	0.1445
Diabetes	0.9273	0.3356
Prior transplants	0.0111	0.9162
Blood units	0.4764	0.4901
Mismatch score	0.5185	0.4715
ALG	10.6502	0.0011

These indicate that ALG is the most significant variable; thus:
Step 1: Variable ALG is entered.
Analysis of variables not in the model:

Variable	LR χ^2	p Value
Age	8.6360	0.0033
Sex	0.0080	0.9287
Dialysis duration	1.9410	0.1636
Diabetes	1.7958	0.1802
Prior transplants	0.4991	0.4799
Blood units	0.4996	0.4797
Mismatch score	0.8352	0.3608

MULTIPLE REGRESSION

Step 2: Variable AGE is entered.
Analysis of variables not in the model:

Variable	LR χ^2	p Value
Sex	0.0179	0.8935
Dialysis duration	0.5319	0.4658
Diabetes	2.8895	0.0892
Prior transplants	0.2179	0.6406
Blood units	0.5441	0.4608
Mismatch score	0.7759	0.3784

Step 3: Variable DIABETES is entered.
Analysis of variables not in the model:

Variable	LR χ^2	p Value
Sex	0.0121	0.9124
Dialysis duration	0.7331	0.3919
Prior transplants	0.1375	0.7108
Blood units	0.7159	0.3975
Mismatch score	0.9845	0.3211

No (additional) variables met the 0.1 level for entry into the model.
Testing the global null hypothesis:

(i) Likelihood test:

$$\chi^2_{LR} = 20.747 \quad \text{with} \quad 3 \text{ dfs} \quad p = .0001$$

(ii) Wald test:

$$\chi^2_W = 21.376 \quad \text{with} \quad 3 \text{ dfs} \quad p = .0001$$

(iii) Score test:

$$\chi^2_S = 21.721 \quad \text{with} \quad 3 \text{ dfs} \quad p = .0001$$

Analysis of maximum likelihood estimates:

Factor	Coefficient	St. Error	z Statistic	p Value
ALG	−0.58475	0.16708	−3.499	0.0005
Age	0.01827	0.00591	3.088	0.0020
Diabetes	0.32304	0.19083	1.693	0.0905

Note: An SAS program would include these instructions:

PROC PHREG DATA=KIDNEY;
MODEL MONTHS*FAILURE(0)=AGE SEX DIALYSIS DIABETES
 PRIORTX BLOOD MISMATCH ALG/
 SELECTION=STEPWISE SLENTRY=.10 SLSTAY=.15 DETAILS;

where KIDNEY is the name assigned to the data set, MONTHS is the variable name for duration time, FAILURE the variable name for survival status of the new graft, and "0" is the coding for censoring. Names listed in the model statement are those assigned to the eight covariates; for example, PRIORTX is the number of prior transplants.

Estimation of the Baseline Survival Function

Consider the proportional hazards model:

$$\lambda[t \mid \mathbf{X} = (x_1, x_2, \ldots, x_k)] = \lim_{\delta \downarrow 0} \frac{\Pr[t \leq T \leq t + \delta \mid t \leq T, \mathbf{X} = \mathbf{x}]}{\delta}$$

$$= \lambda_0(t) e^{[\beta^T \mathbf{x}]}$$

$$= \lambda_0(t) e^{[\beta_1 x_1 + \beta_2 x_2 + \cdots + \beta_k x_k]}$$

which can be expressed equivalently as:

$$S(t \mid \mathbf{X}) = [S_0(t)]^{\exp(\beta^T \mathbf{x})}$$

Suppose now that data are available and we consider the calculation of the nonparametric maximum-likelihood estimate of $S_0(t)$. In doing this, an approach analogous to that used in obtaining the Kaplan–Meier estimate of Chapter 2 is employed.

As before, let

$$t_1 < t_2 < \cdots < t_m$$

Denote the ordered distinct death times. Let R_i be the risk set just before time t_i, D_i the death set at time t_i, and A_i be the set of labels associated with individuals censored in the interval $[t_i, t_{i+1})$. The contribution to the likelihood of an individual with covariates x who died at t_i is:

$$S_0(t_i)^{\exp(\beta^T x)} - S_0(t_i^+)^{\exp(\beta^T x)}$$

and the contribution of a censored observation at time t is:

$$S_0(t^+)^{\exp(\beta^T x)}$$

Therefore, the likelihood function can be written as:

$$L = \prod_{i=0}^{m} \prod_{D_i} [S_0(t_i)^{\exp(\beta^T x)} - S_0(t_i^+)^{\exp(\beta^T x)}] \prod_{A_i} [S_0(t^+)^{\exp(\beta^T x)}]$$

where D_0 is empty.

As with the Kaplan–Meier estimate, consider $S_0(t)$ as a discrete function with probability mass at each t_i; we have:

$$S_0(t) = \prod_{t_i \leq t} (1 - \lambda_i) = \prod_{t_i \leq t} \alpha_i$$

On substitution and rearranging terms, we obtain:

$$L = \prod_{i=0}^{m} \prod_{D_i} [1 - \alpha_i^{\exp(\beta^T x)}] \prod_{R_i - D_i} [\alpha_i^{\exp(\beta^T x)}]$$

as the likelihood function to be maximized.

There are two approaches to this problem:

1. The estimation of the survival function can be carried out by joint estimation of the α's and β. This method also yields an

alternative estimate of β which is close but not identical to the one obtained by using the partial-likelihood function.
2. Most standard computer program, however, implement a more simple solution; that they first estimate β by $\hat{\beta}$ using the partial-likelihood function and then maximize the above likelihood with β replaced by $\hat{\beta}$.

After the estimates $\hat{\alpha}$'s are obtained, the maximum-likelihood estimate of the baseline survival function is then:

$$\widehat{S_0(t)} = \prod_{t_i \leq t} \hat{\alpha}_i$$

and

$$\widehat{S(t|\mathbf{X})} = \prod_{t_i \leq t} \hat{\alpha}_i^{\exp(\hat{\beta}^T \mathbf{x})}$$

Users can specify individual values of all covariates or, if preferred, obtain the survival function estimate for the sample means of these explanatory variables. An example will be given in the next section after an introduction of a simple test of goodness of fit; this will show how to take advantage of the method in order to improve the estimation of a survival curve.

4.3. TIME-DEPENDENT COVARIATES

In prospective studies, since subjects are followed over time, values of many independent variables or covariates may be changing; covariates such as patient age, blood pressure, even treatment. In general, covariates are divided into two categories: *fixed* and *time-dependent*. A covariate is time-dependent if the *difference* between covariate values from two different subjects may be changing with time. For example, sex and age are fixed covariates; a patient's age is increasing by one a year, but the difference in age between two patients remains unchanged. On the other hand, blood pressure is an obvious time-dependent covariate. The following are three important groups of time-dependent covariates.

Examples

(i) Personal characteristics whose measurements are periodically made during the course of a study. Blood pressure fluctuates; so do cholesterol level and weight. Smoking and alcohol consumption habits may change.

(ii) Cumulative exposure: In many studies, exposures such as smoking are often dichotomized; subjects are classified as exposed or unexposed. But this may oversimplified leading to loss of information; the length of exposure may be important. As time goes on, a nonsmoker remains a nonsmoker but "years of smoking" for a smoker increases.

(iii) Another important group are "switching treatments." In a clinical trial, a patient may be transferred from a treatment to another due to side effects or even by patient's request. Organ transplants form another category with switching treatments; when a suitable donor is found, a subject is switched from nontransplanted group to transplanted group. The case of intensive care units is even more complicated where a patient may be moving in and out more than once.

Implementation

Recall that in the analysis using the proportional hazards model we order the death times and form the partial-likelihood function:

$$L = \prod_{i=1}^{m} \Pr(d_i \mid R_i, d_i)$$

$$= \prod_{i=1}^{m} \frac{\exp[\sum_{j=1}^{k} \beta_j s_{ji}]}{\sum_{C_i} \exp[\sum_{j=1}^{k} \beta_j s_{ju}]}$$

where

$$s_{ji} = \sum_{l \in D_i} x_{jl}$$

$$s_{ju} = \sum_{l \in D_u} x_{jl} \qquad D_u \in C_i$$

where R_i is the risk set just before time t_i, n_i the number of subjects in R_i, D_i the death set at time t_i, d_i the number of subjects (i.e., deaths) in D_i, and C_i the collection of all possible combinations of subjects from R_i. In this approach, we try to "explain" why *event(s)* occurred to *subject(s)* in D_i while all subjects in R_i are equally at risk; *this explanation, through the use of* s_{ji} *and* s_{ju}, *is based on the covariate values measured at time* t_i. Therefore, this needs some modification in the presence of time-dependent covariates because events at time t_i should be explained by *values of covariates measured at that particular moment*. Blood pressure, for example, measured years before may become irrelevant.

First, notations are expanded to handle time-dependent covariates. Let x_{jil} be the value of factor x_j measured from individual 1 at time t_i, then the above likelihood function becomes:

$$L = \prod_{i=1}^{m} \Pr(d_i \mid R_i, d_i)$$

$$= \prod_{i=1}^{m} \frac{\exp[\sum_{j=1}^{k} \beta_j s_{jii}]}{\sum_{C_i} \exp[\sum_{j=1}^{k} \beta_j s_{jiu}]}$$

where

$$s_{jii} = \sum_{l \in D_i} x_{jil}$$

$$s_{jiu} = \sum_{l \in D_u} x_{jil} \qquad D_u \in C_i$$

From this new likelihood function, applications of subsequent steps (estimation of β's, formation of test statistics, and the estimation of the baseline survival function) are straightforward; e.g. see Crowley and Hu, 1977. In practical implementation, most standard computer programs have somewhat different procedures for two categories of time-dependent covariates: those that can be defined by a mathematical equation (external) and those measured directly from patients (internal); the former categories are much easier implemented.

Simple Test of Goodness of Fit

Treatment of time-dependent covariates leads to a simple test of goodness of fit. Consider the case of a fixed covariate, denoted by X_1. Instead of the basic proportional hazards model:

$$\lambda(t \mid X_1 = x_1) = \lim_{\delta \downarrow 0} \frac{\Pr[t \leq T \leq t + \delta \mid t \leq T, X_1 = x_1]}{\delta}$$

$$= \lambda_0(t) e^{\beta x_1}$$

we can define an additional time-dependent covariate X_2,

$$\boxed{X_2 = X_1 t}$$

Consider the expanded model:

$$\lambda(t; X_1 = x_1) = \lambda_0(t) e^{\beta_1 x_1 + \beta_2 x_2}$$
$$= \lambda_0(t) e^{\beta_1 x_1 + \beta_2 x_1 t}$$

and examine the significance of

$$\mathcal{H}_0 : \beta_2 = 0$$

The reason for interest in testing whether or not $\beta_2 = 0$ is that $\beta_2 = 0$ implies a goodness of fit of the proportional hazards model for the factor under investigation, X_1. Of course, in defining the new covariate X_2, t could be replaced by any function of t; a commonly used one is:

$$\boxed{X_2 = X_1 \log(t)}$$

This simple approach results in a test of a specific alternative to the proportionality. The computational implementation here is very much similar to the case of cumulative exposures; however, X_1 may be binary or continuous. We may even investigate the goodness of fit for several variables simultaneously. More ideas for general tests for a global alternative will be briefly discussed in the next section.

Example 14. Refer to the data set in Example 1 of Chapter 2 where the remission times of 42 patients with acute leukemia were reported from a clinical trial undertaken to assess the ability of a drug called 6–mercaptopurine (6–MP) to maintain remission. Each patient was randomized to receive either 6–MP or placebo. The study was terminated after one year; patients have different follow-up times because they were enrolled sequentially at different times. Times in weeks were:

6–MP group: 6, 6, 6, 7, 10, 13, 16, 22, 23, 6^+, 9^+, 10^+, 11^+, 17^+, 19^+, 20^+, 25^+, 32^+, 32^+, 34^+, 35^+

Placebo group: 1, 1, 2, 2, 3, 4, 4, 5, 5, 8, 8, 8, 8, 11, 11, 12, 12, 15, 17, 22, 23

in which a $t+$ denotes a censored observation, that is, the case was censored after t weeks without a relapse. For example, "10+" is a case enrolled 10 weeks before study termination and still remission-free at termination.

Since proportional hazards model is often assumed in the comparison of two survival distributions such as in this example (also see Example 1 of Chapter 3), it is desirable to check it for validity (if the proportionality is rejected, it would lend support to the conclusion that this drug does have some cumulative effects).

Let X_1 be the indicator variable defined by:

$$X_1 = \begin{cases} 0 & \text{if placebo} \\ 1 & \text{if treated by 6–MP} \end{cases}$$

and

$$X_2 = X_1 t$$

representing "treatment weeks" (time t is recorded in weeks). In order to judge the validity of the proportional hazards model with respect to X_1, it is the effect of this newly defined covariate X_2 that we want to investigate.

We fit the following model

$$\lambda[t \mid \mathbf{X} = (x_1, x_2)] = \lambda_0(t) e^{[\beta_1 x_1 + \beta_2 x_2]}$$

TABLE 4.8. Investigation of Goodness of Fit

Factor	Coefficient	St. Error	z Statistic	p Value
X_1	−1.55395	0.81078	−1.917	0.0553
X_2	−0.00747	0.06933	−0.108	0.9142

and from the results shown in Table 4.8 it can be seen that the "accumulation effect" or "lack of fit," represented by X_2 is insignificant; in other words, there is not enough evidence to be concerned about the validity of the proportional hazards model.

Note: An SAS program would include these instructions:

```
PROC PHREG DATA=CANCER;
MODEL WEEKS*RELAPSE(0)=DRUG TESTEE;
TESTEE=DRUG*WEEKS;
```

where WEEKS is the variable name for duration time, RELAPSE the variable name for survival status, "0" is the coding for censoring, DRUG is the 0/1 group indicator (i.e., X_1), and TESTEE is the newly created variable (i.e., X_2).

In the next example, we investigate the goodness of fit of two variables simultaneously (the variables being investigated need not be dichotomous; in this example, one is dichotomous the other is continuous).

Example 15. Refer to the data set on acute myelogenous leukemia of Example 1. Patients were classified into the two groups according to the presence or absence of a morphologic characteristic of white cells. Patients termed AG positive were identified by the presence of Auer rods and/or significant granulature of the leukemic cells in the bone marrow at diagnosis. For the AG-negative patients these factors were absent. The other factor under investigation is white blood cells (WBC). Let

$X_1 = \ln(\text{WBC})$

$X_2 = \text{AG group};$ 0 if negative and 1 if positive

$X_3 = X_1 t$

$X_4 = X_2 t$

TABLE 4.9. Investigation of Goodness of Fit

Factor	Coefficient	St. Error	z Statistic	p Value
X_1	0.32194	0.19455	1.655	0.0980
X_2	−0.62867	0.60842	−1.033	0.3015
X_3	0.00284	0.00471	0.602	0.5471
X_4	−0.02532	0.02367	1.070	0.2848

We fit the following model

$$\lambda[t \mid \mathbf{X} = (x_1, x_2)] = \lambda_0(t) e^{[\beta_1 x_1 + \beta_2 x_2 + \beta_3 x_3 + \beta_4 x_4]}$$

and from the results shown in Table 4.9 it can be seen that the "accumulation effects" or "lack of fit," represented by X_3 and X_4 are insignificant; in other words, there is not enough evidence to be concerned about the validity of the proportional hazards model for either one of the two primary covariates.

Example 16. Refer to the data set in Example 1 of Chapter 2 where the remission times of 42 patients with acute leukemia were reported from a clinical trial undertaken to assess the ability of a drug called 6–MP to maintain remission. Each patient was randomized to receive either 6–MP or placebo. The study was terminated after one year; patients have different follow-up times because they were enrolled sequentially at different times. Times in weeks were:

6–MP group: 6, 6, 6, 7, 10, 13, 16, 22, 23, 6+, 9+, 10+, 11+, 17+, 19+, 20+, 25+, 32+, 32+, 34+, 35+
Placebo group: 1, 1, 2, 2, 3, 4, 4, 5, 5, 8, 8, 8, 8, 11, 11, 12, 12, 15, 17, 22, 23

in which a $t+$ denotes a censored observation, that is, the case was censored after t weeks without a relapse. For example, "10+" is a case enrolled 10 weeks before study termination and still remission free at termination.

Since the lack of fit of the proportional hazards model is very insignificant, we can estimate the baseline survival function under this model. Because the covariate $X_1 = 0$ for the placebo group, this

TABLE 4.10. Results From Two Different Methods

Time (in weeks)	Survival Rate Estimate	
	Baseline	Kaplan–Meier
1	0.92039	0.9048
2	0.84078	0.8095
3	0.80097	0.7619
4	0.72136	0.6667
5	0.64174	0.5714
6	0.53093	
7	0.49720	
8	0.36585	0.3810
10	0.33362	
11	0.27039	0.2857
12	0.20566	0.1905
13	0.17528	
15	0.14682	0.1429
16	0.12050	
17	0.09622	0.0952
22	0.04603	0.0476
23	0.01175	0.0000

baseline survival function is also the survival function of the placebo group. However, the estimate of this baseline survival function is based on *both* samples, it is more smooth than the Kaplan–Meier estimate using the placebo sample alone. From the results, shown in Table 4.10 it can be seen that the "baseline" estimate has "jumps" at times where there were events in the *other* group; for example, at $t = 6$.

Note: Computationally, this is a more intensive task than the previous ones. SAS programs, for example, would allow users to obtain an estimate of $S(t \mid \mathbf{X})$ using the relationship:

$$S(t \mid \mathbf{X}) = [S_0(t)]^{\exp(\beta^T \mathbf{x})}$$

but users have to provide values of covariates $\mathbf{X}$. A sample program woud include these instructions:

```
DATA COVAR;
INPUT DRUG;
```

```
CARDS;
0
;
DATA CANCER;
INPUT WEEKS RELAPSE DRUG;
PROC PHREG DATA=CANCER;
MODEL WEEKS*RELAPSE(0)=DRUG;
BASELINE COVARIATES=COVAR OUT=PRED;
PROC PRINT DATA=PRED;
```

where COVAR is the input file specifying covariate values (drug is set at 0 so as to obtain the baseline, survival of the placebo group), and PRED is the output file containing estimates of survival propabilities at various time points.

More on Tests of Goodness of Fit

The proportional hazards model (PHM) expresses a log-linear relationship between X and the hazard function of T:

$$\lambda(t \mid X = x) = \lim_{\delta \downarrow 0} \frac{\Pr[t \leq T \leq t + \delta \mid t \leq T, X = x]}{\delta}$$

$$= \lambda_0(t) e^{\beta x}$$

In the previous section, this proportional hazards assumption was tested by including an additional covariate representing the product of the survival time t (or a function of it) and the covariate under investigation. The test of significance of the newly defined covariate leads us to conclude whether the proportional hazards assumption is valid. This results, of course, in a test of a specific alternative to the proportional hazards assumption.

In addition to the above-mentioned simple suggestion by Cox where one includes one additional covariate representing the product of the survival time t and the covariate under investigation, there are other more complicated methods for testing the validity of the proportional hazards model. Some, such as the approach by Schoenfeld (1980), involve the partition of the time axis. This ad hoc partitioning can be greatly effective if we knew a priori where to look for the period of lack of fit so as to construct an optimal partition of the time axis. The method of Le and Zelterman (1992) is based on the test of central mixtures leading to the following test statistic:

$$S = \sum_{i=1}^{m}[(s_i - \mu_i)^2 - \sigma_i^2]$$

where μ_i and σ_i^2 are the mean and variance of s values over C_i as defined in Section 4.1, that is,

$$\mu_i = \sum_{C_i} s_m/N_i$$

and

$$\sigma_i^2 = \sum_{C_i} s_m^2/N_i - \mu_i^2$$

In words, the test statistic S compares the observed variance $\sum(s_i - \mu_i)^2$ with the weighted true variance $\sum \sigma_i^2$. If the observed variance is much greater than the (estimated) true variance, we can attribute this excess variability to the nonconstant value of the hazards ratio. The test Nagelkerke et al. (1984) was constructed similarly.

Another approach that has become increasingly popular is the analysis of residuals. This graphical method of Therneau, Grambsch, and Fleming (1990) is very useful for investigating the functional form of a covariate as well as the proportional hazards assumption. It starts with a class of martingale-based residuals $M(t)$; this residual measures, at each time t, the difference over $[0,t]$ between the observed number of events and the expected number given the model. For a Cox model with no time-dependent covariates, where $y = (t_i, \delta_i)$ denotes observed data for subject i, this martingale residual reduces to a simple form:

$$\hat{M}_i = \delta_i - \hat{H}_0(t_i)e^{[\hat{\beta}^T x]}$$

in which the baseline cumulative hazards $H_0(t_i)$

$$H_0(t_i) = \int_0^{t_i} \lambda_0(x)\,dx$$

is estimated by

$$\hat{H}_0(t_i) = -\log \hat{S}_0(t_i)$$

using the estimated baseline survival function $\hat{S}_0(t)$. One deficiency of the martingale residual $\hat{M}_i$, particularly in the single event setting of Cox's model, is its skewness. In this setting, it has maximum value $+1$ but a minimum (theoretical) value of $-\infty$. Transformation to achieve a more normal shaped distribution is desirable, particularly when the accuracy of predictions for individual subjects is to be assessed. One such desirable transformation leads to the deviance residuals. For the Cox proportional hazards model, the deviance residuals are:

$$d_i = \text{sgn}(\hat{M}_i) - 2[\hat{M}_i + \delta_i \log(\delta_i - \hat{M}_i)]^{1/2}$$

We now briefly discuss the use of these residuals to assess functional form as well as the validity of the proportional hazards assumption; packaged programs such as SAS provide both types of residuals, but results involving martingale residuals are more difficult to interpret due to its skewness.

Example 17. Refer to the data for patients with acute myelogenous leukemia in Example 1 and suppose we want to investigate the relationship between survival time of AG-positive patients and white blood count (WBC) in two different ways using either (i) $X = $ WBC or (ii) $X = \log(\text{WBC})$. Example 2 indicated that the results would be different for two different choices of X, and this causes an obvious problem of choosing an appropriate measurement scale. Of course, we assume a *linear* model and one choice of X would fit better than the other. Using this new residual analysis, we have

1. For $X = $ WBC, graph of deviance residual versus X is shown in Figure 4.1, from which a lack of fit is rather obvious. Most of the residuals are positive except for a few points with low values of WBC.
2. For $X = \log(\text{WBC})$, graph of deviance residual versus X is shown in Figure 4.2 which shows no indication of a lack of

TIME-DEPENDENT COVARIATES

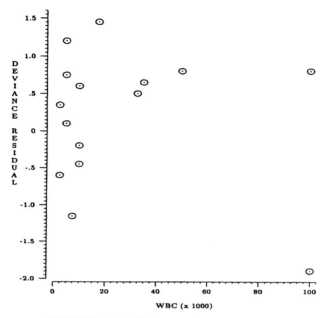

FIGURE 4.1. WBC is used as a covariate.

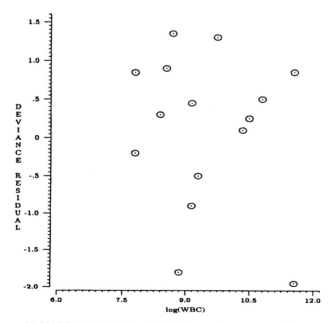

FIGURE 4.2. The log(WBC) is used as a covariate.

fit of the model. In other words, the use of WBC in log scale is justified and preferred.

(If the fit of a group of variables is investigated simultaneously, one would graph the deviance residual versus the *combined linear predictor*:

$$\beta^T \mathbf{x} = \beta_1 x_1 + \beta_2 x_2 + \cdots + \beta_k x_k$$

However, it is not possible to single out the variable or variables responsible for any lack of fit.)

Note: An SAS program would include these instructions:

```
INPUT WBC WEEKS DEATH;
PROC PHREG DATA=CANCER;
MODEL WEEKS*DEATH(0)=WBC;
OUTPUT OUT=TEMP RESDEV=DEV;
PROC PLOT DATA=TEMP;
PLOT DEV*WBC='*';
```

where CANCER is the name assigned to the data set, WEEKS is the variable name for duration time, DEATH the variable name for survival status, and "0" is the coding for censoring. TEMP is the name of the temporary output file containing deviance residuals and symbol '*' is chosen to mark the points on the graph.

Example 18. Refer to the data set in Example 1 of Chapter 2 where the remission times of 42 patients with acute leukemia were reported from a clinical trial undertaken to assess the ability of 6–MP to maintain remission. In Example 14, the proportional hazards assumption was tested by including an additional covariate representing the product of the survival time t and drug indicator, the covariate under investigation. The result showed that the lack of fit of the proportional hazards model is very insignificant. However, when the model including only the original drug indicator was fitted, Figure 4.3 shows a mild pattern of deviations from proportionality; the same result was also found using the test of Le and Zelterman (1992).

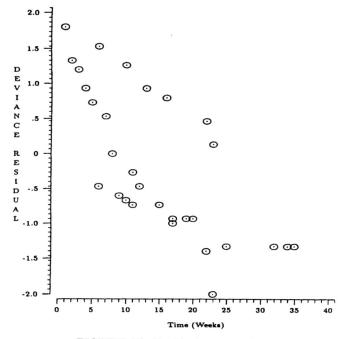

FIGURE 4.3. Residuals versus time.

4.4. STRATIFICATION AND ITS APPLICATIONS

The proportional hazards model requires that, for a covariate X—say an exposure—the hazards functions at different levels, $\lambda(t;\text{exposed})$ and $\lambda(t;\text{nonexposed})$ are proportional. Of course, sometimes there are factors, the different levels of which produce hazard functions that deviate markedly from proportionality. These factors may not be under investigation themselves, especially those of no intrinsic interest, those with a large number of levels and/or those where interventions are not possible. But these factors may act as important confounders that must be included in any meaningful analysis so as to improve predictions concerning other covariates. Common examples include sex, age, neighborhhood, and the like. To accommodate such confounders, an extension of the proportional hazards model is desirable. In this section, we will consider such an extension and then study a special case, its application to the analysis of case-control studies with multiple matching.

Basic Ideas

Suppose there is a factor that occurs on q levels and for which the proportional hazards model may be violated. If this factor is under investigation as a *covariate*, then the model and subsequent analyses are not applicable. However, if this factor is *not* under investigation and is considered only as a confounder so as to improve analyses and/or predictions concerning other covariates, then *we can treat it as a stratification factor*. By doing that we will get no results concerning this factor (which are not wanted), but in return we do not have to assume that the hazard functions corresponding to different levels are proportional (which may be severely violated). Since many stratification factors can be combined into one, we will present the model only for the case of one stratification factor.

Model and Implementation

Again, suppose the stratification factor Z has q levels; this factor is not clinically important itself, but adjustments are still needed in efforts to investigate other covariates. We define the hazard function for an individual in the jth stratum (or level) of this factor as:

$$\lambda[t \mid \mathbf{X} = (x_1, x_2, \ldots, x_k)] = \lambda_{0j}(t) e^{[\beta^T \mathbf{x}]}$$
$$= \lambda_{0j}(t) e^{[\beta_1 x_1 + \beta_2 x_2 + \cdots + \beta_k x_k]}$$

for $j = 1, 2, \ldots, q$ where $\lambda_0(t)$ is an *unspecified* baseline hazard for the jth stratum and $\mathbf{X}$ representing other k covariates under investigation (excluding the stratification itself). The baseline hazard functions are allowed to be arbitrary and are completely unrelated (and, of course, *not* proportional). The basic additional assumption here, which is the same as that in the *analysis of covariance*, requires that the β's are the same across strata (i.e. the so-called parallel lines assumption, which is testable).

In the analysis, we identify distinct times of events for the jth stratum and form the partial-likelihood $L_j(\beta)$ as in the previous sections. The *overall* partial likelihood of β is then the product of those q stratum-specific likelihoods:

$$L(\beta) = \prod_{j=1}^{q} (\beta)$$

Subsequent analyses, finding maximum-likelihood estimates as well as using score statistics, are straightforward. For example, if the null hypothesis $\beta = 0$ for a given covariate is of interest, the score approach would produce a *stratified* log-rank test.

The division into strata does not need to assume the proportional hazards model for the factor concerned but also tends to reduce the computational complexities of partial-likelihood inference by reducing the number of ties. Once the estimate of β is obtained, the method of Section 4.2 can be applied to estimate the stratum-specific baseline survival functions $S_{0j}(t)$, $j = 1, 2, \ldots, q$, which can be use graphically to check whether the factor itself can be modeled under the proportional hazards assumption.

The next subsection introduces an application of stratification to a popular epidemiological retrospective design.

Analysis of Epidemiologic Matched Studies

When comparing disease cases to controls (subjects without the disease under investigation), investigators are often intestested in evaluating the effects of potential risk factors on the disease in terms of odds ratios. As a technique for the control of confounding factors in a designed study, individual cases are matched to a set of (one or more) controls chosen to have similar values for the important confounding variables (which are not under investigation themselves). One-to-one matching is the most popular and most cost-effective design. However, when control subjects are more readily acquired than cases—which is almost always true—it is surely more efficient to match more than one control to each case. In the situation where there is a fixed number R of controls matched to each case and only one dichotomous risk factor is under investigation, the solution is simple (see Breslow and Day, 1980).

In the situation where there are several potential risk factors, some of which may be continous, and/or a variable number of cases matched to each control, a logical approach is through the use of logistic regression. For matched studies, the use of conditional logistic regression model is now widely used; a good review of the method is given by Breslow (1982). For the general case of n_i cases matched to m_i controls in a set, we have the conditional probability of the observed outcome (that a specific set of n_i subjects are cases) given that the number of cases is n_i (any n_i subjects could be cases):

$$\frac{\exp(\sum_{j=1}^{n_i}[\beta^T \mathbf{x_j}])}{\sum_{R(n_i,m_i)} \exp(\sum_{j=1}^{n_i}[\beta^T \mathbf{x_j}])}$$

where the sum in the denominator ranges over the collections $R(n_i, m_i)$ of all partitions of the $(n_i + m_i)$ subjects into two, one of size n_i and one of size m_i. The full conditional likelihood is the product over all matched sets; one probability for each set.

From the expression of the product of the above probabilities, it can be seen the full likelihood of the observed outcome has the same mathematical form as the overall partial likelihood for the proportional hazards survival model with strata, one for each matched set, and one event time for each. This enables us to adapt programs written for proportional hazards model to analyze epidemiologic matched studies. The essential features of the adaptation are:

(i) Creating matched set numbers and using them as different levels of a stratification factor.

(ii) Assigning a number to each subject; these numbers will be used in the place of duration times. These numbers are chosen arbitrarily as long as the number assigned to a case is smaller than the number assigned to a control in the same matched set. This is possible because when there is only one event in each set, the numerical value for the time to event becomes irrelevant.

Example 19. Consider the data taken from a study by Herbst et al. (1971); the cases were eight subjects 15 to 22 years of age who were diagnosed with vaginal carcinoma between 1966 and 1969. For each case, four controls were found in the birth records of patients born within 5 days of the case in the same hospital. The risk factors of interest are the mother's bleeding in this pregnancy (N = No, Y = Yes) and any previous pregnancy loss by the mother (N = No, Y = Yes). The data are given in Table 4.11, as used by Holford (1982). With the usual coding 0/1 (0 = No, 1 = Yes) for exposure and case-control (0 = control, 1 = case), a made-up time variable could be defined as (2-case) (so that time = 1 for a case,

TABLE 4.11. Data from a One-to-Four Matched Study

		Responses (Bleeding, Previous Loss) Control Subject Number			
Set	Case	1	2	3	4
1	(N,Y)	(N,Y)	(N,N)	(N,N)	(N,N)
2	(N,Y)	(N,Y)	(N,N)	(N,N)	(N,N)
3	(Y,N)	(N,Y)	(N,N)	(N,N)	(N,N)
4	(Y,Y)	(N,N)	(N,N)	(N,N)	(N,N)
5	(N,N)	(Y,Y)	(N,N)	(N,N)	(N,N)
6	(Y,Y)	(N,N)	(N,N)	(N,N)	(N,N)
7	(N,Y)	(N,Y)	(N,N)	(N,N)	(N,N)
8	(N,Y)	(N,N)	(N,N)	(N,N)	(N,N)

TABLE 4.12. Results for Data in Table 4.11

Factor	Coefficient	St. Error	z Statistic	p Value
Bleeding	1.61977	1.36892	1.183	0.2367
Previous Loss	1.73186	0.89340	1.938	0.0526

and = 2 for a control). The results, shown in Table 4.12, indicate that the previous pregnancy loss is almost significant at the 5 percent level ($p = .0526$) but the factor "bleeding" is not statistically significant ($p = .2367$).

Note: An SAS program would include these instructions:

```
INPUT SET CASE BLEEDING PLOSS;
DUMBTIME=2-CASE;
PROC PHREG DATA=CANCER;
MODEL DUMBTIME*CASE(0)=BLEEDING PLOSS;
STRATA SET;
```

where SET number goes from 1 to 8, CASE is the disease indicator (= 1 for a case and 0 for a control), BLEEDING and PLOSS (for Pregnancy loss) are the two risk factors, and DUMBTIME is the newly created variable used as duration time.

Other analyses with stratification are implemented similarly. For example, in Example 13 (and Example 9), if we decided to treat DIABETES as a stratification factor—instead of a covariate—then the SAS program would be modified as follows:

```
PROC PHREG DATA=KIDNEY;
MODEL MONTHS*FAILURE(0)=AGE SEX DIALYSIS PRIORTX
  BLOOD MISMATCH ALG;
STRATA DIABETES;
SELECTION=STEPWISE SLENTRY=.10 SLSTAY=.15 DETAILS;
```

The analysis of a 1-to-R matched study involving only one dichotomous turns out quite simple. Let n_m of matched sets with m exposed individuals (case and/or controls), $0 \le m \le (R+1)$. Then we can decompose:

$$n_m = n_{1,m-1} + n_{0,m}$$

where $n_{1,m-1}$ is the number of sets with m-exposed individuals and with an exposed case whereas $n_{0,m}$ is the number of sets m-exposed individuals and an unexposed case. The data can be condensely and sufficiently presented as shown in Table 4.13. (The sets with $m = 0$ and $n = R + 1$ do not contribute.) It can be shown that the (stratified) z statistics of Section 4.1,

$$z = \left\{ \sum_{i=1}^{R}(s_i - \mu_i) \right\} \bigg/ \left\{ \sum_{i=1}^{R} \sigma_i^2 \right\}^{1/2}$$

is reduced to:

$$z = \left\{ \sum_{i=1}^{R}\left(n_{1,m-1} - \frac{mn_m}{R+1}\right) \right\} \bigg/ \left\{ \frac{\sum_{i=1}^{R} m(R-m+1)n_m}{(R+1)^2} \right\}^{1/2}$$

TABLE 4.13. Data from a One-to-R Matched Study

Case	Number of Exposed Individuals m				
	1	2	...	m	...	R
Exposed	$n_{1,0}$	$n_{1,1}$	...	$n_{1,m-1}$	...	$n_{1,R-1}$
Unexposed	$n_{0,1}$	$n_{0,2}$	...	$n_{0,m}$	...	$n_{0,R}$

which is also often used in the form of a chi square with one degree of freedom. The maximum-likelihood estimate of the odds ratio is still obtained iteratively but another estimate, the *Mantel–Haenszel* estimate, which is also popular:

$$\hat{\theta}_{MH} = \left\{\sum_{i=1}^{R}(R - m + 1)(n_{1,m-1})\right\} \Big/ \left\{\sum_{i=1}^{R} mn_{0,m}\right\}^{1/2}$$

Example 20. Breslow and Day (1980) used the data on endometrial cancer which were taken from Mack et al. (1976). The investigators identified 63 cases of endometrial cancer occurring in a retirement community near Los Angeles, California, from 1971 to 1975. Each disease individual was matched with $R = 4$ controls who were alive and living in the community at the time the case was diagnosed, who were born within one year of the case, who were the same marital status, and who had entered the community at approximately the same time. The risk factor was previous use of estrogen. The data are given in Table 4.14. The value of the z statistics is $z = 5.582$ or, equivalently, $X^2 = 31.156$, which is of course highly significant. The maximum-likelihood estimate of the odds ratio is:

$$\hat{\theta} = 7.95$$

and the corresponding Mantel–Haenszel estimate is:

$$\hat{\theta}_{MH} = 8.46$$

TABLE 4.14. Data from a Study of Endometrial Cancer

Case	No. Exposed m			
	1	2	3	4
Exposed	3	17	16	15
Unexposed	4	1	1	1
Total	7	18	17	16

Homogeneity of Odds Ratio in Matched Studies

The final remarks are concerned with an important aspect of case-control studies, the homogeneity of the odds ratio. The 1-to-R matched design has been used often in epidemiological studies and has proved useful and economical when disease individuals, called cases, are rare. This design can be analyzed as a special case of survival data with stratification where each matched set serves as a stratum and, hence, its only risk set (with one event, the case). The method of Le and Zelterman (1992) for goodness of fit can be used to test for the homogeneity of the odds ratio:

$$S = \sum_{i=1}^{m}[(s_i - \mu_i)^2 - \sigma_i^2] \tag{8}$$

In this simple case it can be shown that

$$\mu_i = p_m$$
$$= \frac{me^\beta}{me^\beta + R - m + 1}$$

where p_m, $0 \leq m \leq (R+1)$ is the probability that the case has been exposed given that there are m-exposed members of a set. In addition,

$$\sigma_i^2 = p_m q_m$$

with $q_m = 1 - p_m$, leading to the test statistic:

$$S = \sum_{m=1}^{R} p_m(n_m p_m - n_{1,m-1})$$

If β is replaced by $\hat{\beta}$ in p_m and q_m, yielding $\hat{p}_m$ and $\hat{q}_m$, then the estimated variance of $\hat{S}$ is given by:

$$\widehat{\mathrm{Var}}(\hat{S}) = \sum_{m=1}^{R} n_m \hat{p}_m^3 \hat{q}_m - \frac{[\sum_{m=1}^{R} n_m \hat{p}_m^2 \hat{q}_m]^2}{\sum_{m=1}^{R} n_m \hat{p}_m \hat{q}_m}$$

The standardized statistic $\hat{S}/\text{Var}(\hat{S})^{1/2}$, which is distributed asymptotically as standard normal, is identical to the statistic T_5 of Liang and Self (1985); see Zelterman and Le (1991) and Jones et al. (1989) for more details on these tests of homogeneity.

4.5. A NONTIME APPLICATION: EVALUATION OF CONFOUNDING EFFECTS IN ROC STUDIES

Survival analysis is in fact a loosely defined statistical term that encompasses a variety of statistical techniques for analyzing positive-valued random variables. As we have seen, typically the value of the random vaiable is the *time* to the failure of a physical component (mechanical or electrical) or the *time* to the death of a bioligical unit (patient, animal, cell, etc.). However, it could be the *time* to the learning of a skill or it *may not even be time at all*. This section provides a special application where the random variable is not time but another well-defined positive-valued quantity. This application outlines the use of Cox's proportional hazards regression for the evaluation of confounding effects in the receiver operating characteristic (ROC) studies (Le, 1997). The first part reviews the background of this area.

In epidemiological studies much use is made of diagnostic tests, based either on clinical observations or on laboratory techniques, by means of which individuals are classified as healthy or as falling into one of a number of disease categories. Such tests are, of course, important throughout the whole of medicine and public health, and in particular form the basis of screening programs for the early diagnosis of disease. It is often true that the result of the test, although dichotomous, is based on the dichotomization of a continuous variable—say, X—herein referred to as the *separator variable*. Let us assume without loss of generality that smaller values of X are associated with the diseased population, often called the *population of the cases*. Conversely, larger values of the separator are assumed to be associated with the control or nondiseased population.

A test result is classified by choosing a cut-off $X = x$ against which the observation of the separator is compared. A test result is positive if the value of the separator does not exceed the cut-off; otherwise, the result is classified as negative. Most diagnostic tests are imperfect instruments, in the sense that healthy individuals will occasionally be classified wrongly as being ill, while some individuals who are really ill may fail to be detected as such. Therefore, there is the ensuing conditional probability of the correct classification of a randomly selected case, called the *sensitivity* of the test:

$$F(x) = \Pr(X \leq x \mid \text{case})$$

This is the *cumulative distribution function* (cdf) of the separator variable for the cases. Similarly, the conditional probability of the correct classification of a randomly selected control, $\Pr(X > x \mid \text{control})$, is called the *specificity* of the test. It is obvious that

$$G(x) = 1 - \text{specificity}$$
$$= \Pr(X \leq x \mid \text{control})$$

is the cdf of the separator variable for the controls. A *receiver operating characteristic* (ROC) curve, the trace of the sensitivity versus (1 − specificity) of the test, is generated as the cut-off x moves through its range of possible values. In other words, the ROC function R associated with a separator X is defined by

$$\boxed{F(x) = R[G(x)]}$$

Estimator for the ROC Function

Given two independent samples, $\{x_{1i}\}_{i=1}^{m}$ and $\{x_{2j}\}_{j=1}^{n}$, from m controls and n cases, respectively. Let R_j be the rank of x_{2j} among the cases $\{x_{2j}\}_{j=1}^{n}$ ($1 \leq R_j \leq n$); S_j be the rank of x_{2j} among the pooled sample $\{x_{1i}, x_{2j}\}$ ($1 \leq S_j \leq n + m$); and

$$\boxed{u_j = \frac{S_j - R_j}{m} \qquad 1 \leq j \leq n}$$

the percent of controls to the left of case j. Then the step function R_n is defined by

$$\boxed{R_n(u) = (\text{number of } u_j \leq u)/n \qquad 0 \leq u \leq 1}$$

converges uniformly to R (Hoeffding, 1954; Kadane, 1971; Le, 1997). As an important implication of our result, where we can estimate the ROC function R by R_n, we can treat $\{u_i\}_{i=1}^n$ as a *pseudo-sample* from a population distributed with cdf R.

In recent years, diagnostic research has begun to address the potentialities and the problems associated with the incorporation of concomitant information into ROC studies. For example, Ahlquist et al. (1985), when studying the diagnostic performance of a dye test for the detection of fecal blood, found that its specificity and sensitivity were markedly affected by the condition of the sample material itself (e.g., wetness, temperature, etc.). Similarly, when studying the diagnostic performances of admittance and tympanometric width for the detection of middle ear fluid (Le et al., 1992), we found that their diagnostic powers were affected by otoscopic findings. Hence, researchers often look for methods for the inclusion of concomitant information when evaluating diagnostic tests (Hlatky et al., 1984). A possible approach to account for extraneous factors is regression methodology where we can take advantage of the new estimator for the ROC function and use Cox's proportional hazards regression for the evaluation of confounding effects in ROC studies. This method can be used even when concomitant information is only available for the cases, for example, disease severity.

Use of Proportional Hazards Model

After the above preliminary calculations, the data we have are now in the form

$$\{u_i, z_{1i}, z_{2i}, \ldots z_{ki}\}_{i=1}^n$$

where the u_i's are the newly calculated pseudo-sample values and $Z_1, Z_2, \ldots, Z_k$ are k covariates. We now can use as *times* the y_i's defined by

$$y_i = -\ln u_i$$

and apply proportional hazards regression to the sample

$$\{y_i; z_{1i}, z_{2i}, \ldots z_{ki}\}_{i=1}^n$$

(we may have to replace u_i by $1/(2m)$ if $u_i = 0$).

Consider the case of one covariate Z, then it can be shown that the above proportional hazards model can be expressed as:

$$R(u;z) = \varphi(u)^{\exp[\beta z]} \qquad 0 \leq u \leq 1$$

where

$$\varphi(u) = S_0(-\ln u)$$

with $S_0(\cdot)$ being some survival function, that is, nondecreasing, from $(0, \infty)$ to $(0, 1)$. We can easily see the coefficients β's representing the effects of the covariates on the height of the ROC curve. When we fix the specificity $(1 - u)$ of a screening test, $R(u)$ represents the height of the ROC curve or the sensitivity of the procedure, and:

(i) If Z is a binary covariate, coded as 0/1, then

$$\text{Sensitivity}(u, Z = 1) = [\text{Sensitivity}(u, Z = 0)]^{\exp(\beta)}$$

(ii) If Z is a continuous covariate, then

$$\text{Sensitivity}(u, Z = z + 1) = [\text{Sensitivity}(u, Z = z)]^{\exp(\beta)}$$

In this formulation we use only information on the covariates measured from the cases. The aim here is focused more on the effects of concomitant information on sensitivity of the screening procedure. The parameters β and $\exp(\beta)$ may not have an interesting interpretation such as "relative risk" or "odds ratio" in logistic and Cox's regression models; the method is aimed at identifying important covariates. However, on the other hand, the values of the regression coefficients lead, collectively to an important application. Using current statistical methodology associated with Cox's model, we can estimate these coefficients as well as the baseline survival function S_0 [leading to an estimate of $\varphi(u)$]; see Section 4.2. The results are then used to estimate the ROC curve at specific values of covariates, for example,

$$\hat{R}(u; z_1 = z_{10}, z_2 = z_{20}, \ldots)$$

which is obviously relevant to clinical applications.

Special Case

Consider, again, the case of one covariate Z:

$$R(u;z) = \varphi(u)^{\exp[\beta z]} \qquad 0 \leq u \leq 1$$

When the baseline survival is exponential or

$$S_0(y) = \exp(\theta y)$$

then

$$\varphi(u) = \exp[(\theta)(-\ln u)]$$
$$= u^\theta$$

which is often referred to as *the Lehmann alternatives*. That is,

$$R(u) = u^\theta \qquad 0 \leq u \leq 1 \quad \text{and} \quad 0 \leq \theta \leq 1$$

for which the parameter θ characterizes the diagnostic power of the separator representing the area under the ROC curve. The use of this model may have started even before Bamber (1975); also see Hanley and McNeil (1982).

Example 21. As part of ongoing clinical studies of otitis media (ear infection), children ranging in age from 6 months to 8 years from a suburban multispecialty clinic underwent myringotomy and tympanostomy tube placement between March 1987 and January 1990. Enrolled subjects underwent extensive preoperative examination to determine demographic and clinical characteristics; clinical evaluation included otoscopy observations of the tympanic membrane (i.e., color, position, appearance, and mobility) and tympanometric assessment. Among single-frequency (226 Hz) tympanometric measurements the static admittance has emerged as a leading potential separator for the detection of middle ear fluid. The static admittance was determined by $X = (X_{\max} - X_{+200})$

TABLE 4.15. Results of Proportional Hazards Regression Analysis

Factor	Regression Results using Propotional Hazards Model		
	Coeff. β	Std. Error	z Score
Mucosa appearance	0.082	.181	0.45
Position	0.171	.171	1.00
Appearance (TM)	−0.244	.219	−1.11
Color	0.041	.162	0.25
Mobility	−1.704	.365	−4.66
Age	−0.004	.003	−1.07
MEE Viscosity	0.061	.192	0.32

where X_{max} is the peak admittance and X_{+200} is the admittance at 200 da Pa. At myringotomy, the primary outcome measure is the presence (case) or absence (control) of middle ear fluid. Measures of disease severity included mucosa appearance (normal/abnormal, normal being pale and thin), and if fluid was present, viscosity was determined (thin vs. thick or gluelike). Mucosa appearance is observed from all ears, but viscosity is only classified when middle ear fluid is present (i.e., cases). On the average, surgery was 2 to 3 days after otoscopic examination and tympanometric measurements. A total of 247 ears with complete data are included in the analysis; middle ear fluid was found in 185 ears.

All of the previously mentioned suspected risk factors were included in a propotional hazards regression analysis and Table 4.15 gives the corresponding results: values for $\hat{\beta}_i$, its standard error, and the standardized normal score:

$$\hat{\beta}_i/SE(\hat{\beta}_i)$$

These results singled out "tympanic membrane mobility" as the only significant confounder; in fact, "tympanic membrane appearance" was also found significant in an univariate analysis, but its level of significance was greatly reduced because appearance and mobility are greatly correlated.

EXERCISES

4.1. In the log-rank test of Section 4.1,

$$z = \left\{\sum_{i=1}^{k}(s_i - \mu_i)\right\} \Big/ \left\{\sum_{i=1}^{k}\sigma_i^2\right\}^{1/2}$$

we treat D_i, the dead set, as a sample of size one taken from C_i (so that μ_i is the mean and σ_i^2 is the variance of s over C_i—considered as a finite population). An alternative approach is to treat D_i as a sample of size d_i taken from the risk set R_i. Apply this test, with proper use of a correction factor for finite population, to the data set in Example 3; compare the result to that in Example 3.

4.2. There are two nonparametric versions of the log-rank test for a continuous covariate, one is based on *the rank of the sum* and one is based on *the sum of the ranks*. Apply both of these to the data of the large standard group in Example 14 of Chapter 1 in order to investigate the relationship between t and x_2.

4.3. In Example 5 of Chapter 1, data were given for two groups of patients who died of acute myelogenous leukemia. Suppose we want to investigate the relationship between survival time of AG-negative patients and WBC in two different ways using (i) X = WBC then (ii) X = log(WBC).
  (i) Find the relative risk for (WBC = 100,000) vs. (WBC = 50,000).
  (ii) Compare the results to those in Example 2. Is there an effect modification here?

4.4. Appendix A gives the data for 469 patients with kidney transplants; the primary endpoint was graft survival and time to graft failure was recorded in months. Perform a simple regression with only one covariate, the use of ALG (1 if used, 0 otherwise), and calculate a 95 percent confidence interval for the corresponding relative risk; draw your conclusion.

4.5. Refer again to the data set on kidney transplants of Appendix A and Exercise 4.4, but this time we investigate only one covariate,

$$X = (\text{Age})^2$$

Why is the result not the same as that in Example 10?

4.6. Refer to the myelogenous leukemia data of Example 1 and let G be the AG-group indicator; 0 if negative and 1 if positive. Perform a simple regression with only one covariate:

$$X = G \ln(\text{WBC})$$

Compare the result to that concerning "Product" in Example 11. Draw your conclusion.

4.7. Refer to the data set in Example 1 of Chapter 2 where the remission times of 42 patients with acute leukemia were reported from a clinical trial undertaken to assess the ability of 6–MP to maintain remission. Find 95 percent confedence intervals for the relative risk associated with the use of the drug 6–MP using two different coding schemes and compare the results:

$$X_i = \begin{cases} 0 & \text{if placebo} \\ 1 & \text{if drug 6–MP} \end{cases}$$

and

$$X_i = \begin{cases} -1 & \text{if placebo} \\ 1 & \text{if drug 6–MP} \end{cases}$$

4.8. Refer to data for the squamous, standard group of lung cancer patients in Example 14 of Chapter 1. Test to see if, taken collectively, the four covariates (performance status, time from diagnosis, age, and prior therapy) contribute significantly to the prediction of survivorship.

4.9. Appendix A gives the data for 469 patients with kidney transplants; the primary endpoint was graft survival and time to graft failure was recorded in months. Investigate the collective effect of days of hemodialysis, age at transplant, amount of blood transfusion, and the interaction between the last two factors. Is the intraction between age and blood transfusion statistically significant?

4.10. Refer again to the kidney transplants data of Appendix A with the eight covariates investigated in Example 9. Evaluate the additional collective effects of the following three intraction terms:

ALG∗Dialysis duration
ALG∗Mismatch Score
ALG∗Diabetes

Compare the result for the last term with that found in Example 12.

4.11. Refer again to the data set on kidney transplants of Appendix A and repeat the analysis in Example 7 but not including Diabetes as a covariate; however, do it separately for two subsamples: those with diabetes and those without diabetes. "Grossly" compare the slopes for each of the seven covariates; does the results justify the use of Diabetes as a stratification factor? If it seems they do, carry out the stratified analysis and calculate a 95 percent confidence interval for the relative risk associated with the use of ALG.

4.12. Refer to the myelogenous leukemia data of Example 1. Patients were classified into the two groups according to the presence or absence of a morphologic characteristic of white cells and the primary covariate is WBC. Let G be the AG-group indicator; 0 if negative and 1 if positive and perform a multiple regression with two covariates:

$$X_1 = \ln(\text{WBC})$$
$$X_2 = GX_1$$

Does it justify to use the AG group, represented by G, as stratification factor in order to investigate the effect of WBC?

4.13. Refer to the myelogenous leukemia data of Example 1. Perform a simple regression to investigate the effect of ln(WBC) but do it separately for the two groups of patients: AG positives and AG negatives. In each case, obtain the baseline survival curve; graph these curves to determine if they have proportional hazards.

4.14. Refer to the data set in Example 1 of Chapter 2 where the remission times of 42 patients with acute leukemia were reported from a clinical trial undertaken to assess the ability of 6–MP to maintain remission. Graph the two Kaplan–Meier curves, one for each group, to determine if they have proportional hazards; compare the finding to those in Examples 14 and 18.

4.15. Refer again to the data set on kidney transplants of Appendix A with the same eight covariates investigated in Example 9. Perform a stepwise regression using the subsample consisting of those with diabetes. Do we have the same results as in Example 13?

4.16. Refer again to the data set on kidney transplants of Appendix A with age as the only covariate. Form two subgroups of patients: (1) Age $\leq$ 45 and (2) Age $\geq$ 60; for each subgroup, perform a simple regression (age is the covariate). Compare the levels of significance as well as the relative risks and comment on your findings.

4.17. Refer again to the data set on kidney transplants of Appendix A with diabetes and ALG as the only covariates. Test for the validity of the proportional harzards model for both covariates simultaneously.

4.18. Refer to the myelogenous leukemia data of Example 1. Use

$$X_1 = \ln(WBC)$$

$$X_2 = \text{AG group;} \quad 0 \text{ if negative and } 1 \text{ if positive}$$

in a multiple regression analysis to obtain deviance residuals. Graph the residuals versus time and comment on your findings regarding the validity of the proportional hazards assumption.

4.19. Refer to the data set on ovarian carcinoma of Exercise 3.18. Define

$$X = \begin{cases} 0 & \text{for a patient with low grade} \\ 1 & \text{for a patient with high grade} \end{cases}$$

Investigate the validity of the proportional hazards assumption by three different ways:
  (i) Graph the two Kaplan–Meier curves (on loglog scale).
  (ii) Fit a simple regression model and graph the deviance residuals versus time.
  (iii) Fit a multiple regression model including X and an additional covariate representing the product of X and the survival time t.
Comment on your findings.

4.20. Refer to the data on vaginal carcinoma of Example 19 but focus only on the factor Bleeding.
  (i) Put the data into a 2×2 table.
  (ii) Calculate the Mantel–Haenszel estimate of the odds ratio.
  (iii) Test the null hypothesis that factor Bleeding is not related to the disease and compare your result to that in Example 19. Why are they different?

APPENDIX A

Kidney Transplant Data

OBS	AGE	SEX	DIALY	DBT	PTX	BLOOD	MIS	ALG	MONTH	FAIL
1	32	0	105	0	0	14	1	0	12	1
2	40	0	12	0	0	14	1	0	1	1
3	45	0	26	1	0	14	1	0	9	1
4	45	1	0	0	0	14	1	0	1	1
5	22	1	167	0	0	14	1	0	28	1
6	25	0	0	0	0	14	1	0	3	1
7	42	1	0	0	0	14	1	0	1	1
8	24	0	0	0	0	14	1	0	8	1
9	23	0	0	0	0	14	1	0	180	1
10	38	0	0	0	0	14	1	0	3	1
11	32	0	0	0	0	14	1	1	2	1
12	25	1	210	0	1	14	1	0	40	1
13	22	1	82	0	0	20	1	1	5	1
14	42	0	400	0	0	0	1	0	4	1
15	48	0	331	0	0	14	2	1	4	1
16	35	0	625	0	0	14	2	0	15	1
17	27	0	492	0	0	8	0	0	200	0
18	28	0	774	0	0	3	1	0	44	1
19	27	0	749	0	1	14	1	1	4	1
20	31	1	883	0	0	7	0	1	183	0
21	14	0	48	0	0	5	1	1	18	1
22	19	1	97	0	0	17	4	1	13	0
23	36	1	247	0	0	14	3	1	96	1
24	30	1	411	0	0	17	2	1	178	0
25	27	1	470	0	0	61	2	1	177	0
26	29	1	648	0	0	39	2	1	10	1
27	35	0	456	0	0	6	1	1	14	1

KIDNEY TRANSPLANT DATA

OBS	AGE	SEX	DIALY	DBT	PTX	BLOOD	MIS	ALG	MONTH	FAIL
28	40	1	901	0	0	14	1	0	4	1
29	46	1	765	0	0	60	1	1	136	1
30	35	0	1178	0	0	15	2	1	166	0
31	37	1	608	0	0	9	1	1	22	1
32	43	1	60	0	0	4	4	1	10	1
33	30	0	407	0	0	3	2	0	114	1
34	43	1	264	0	0	14	1	1	66	1
35	28	1	156	0	2	17	1	1	158	0
36	16	1	372	0	0	21	1	1	71	1
37	22	0	606	0	0	21	0	1	56	1
38	31	0	109	0	0	20	2	1	155	0
39	47	0	251	0	0	2	2	0	9	1
40	38	1	153	0	0	8	2	1	19	0
41	34	0	1925	0	1	14	2	1	65	1
42	24	0	1078	0	0	32	2	1	153	0
43	42	1	2199	0	0	111	2	1	151	0
44	44	1	76	0	0	4	0	1	137	1
45	29	0	217	0	0	12	0	1	1	1
46	28	1	327	0	0	10	1	1	108	1
47	31	0	1118	0	0	2	2	1	9	1
48	43	0	1720	0	0	53	2	1	8	1
49	27	1	402	0	0	21	3	1	132	1
50	46	0	74	0	0	7	1	1	112	1
51	40	0	42	0	0	0	2	1	51	1
52	43	1	53	0	0	9	1	1	145	0
53	47	0	147	0	0	8	2	1	141	0
54	36	1	529	0	0	20	2	1	2	1
55	24	0	247	0	0	9	2	1	138	0
56	38	0	183	0	0	11	3	1	136	0
57	29	0	424	0	0	20	2	1	22	0
58	34	0	960	0	0	24	0	1	52	1
59	29	0	487	1	0	9	2	1	1	1
60	45	1	444	0	1	117	2	1	6	1
61	43	1	804	0	0	14	3	1	1	1
62	31	1	209	1	0	12	4	1	1	1
63	49	0	469	0	1	2	2	1	2	1
64	35	0	1471	0	0	3	4	1	130	0
65	34	1	523	0	0	7	1	1	2	1
66	19	1	87	0	0	6	2	1	127	0
67	24	0	52	0	0	2	2	1	127	0
68	43	0	190	1	0	2	2	0	2	1
69	20	1	471	0	0	9	1	1	115	1

KIDNEY TRANSPLANT DATA

OBS	AGE	SEX	DIALY	DBT	PTX	BLOOD	MIS	ALG	MONTH	FAIL
70	39	1	235	0	0	0	3	0	2	1
71	50	0	193	0	0	14	0	1	104	1
72	22	0	49	0	0	3	1	0	120	0
73	48	0	294	0	0	6	3	0	4	1
74	21	0	592	0	0	17	3	1	74	1
75	55	0	1284	0	0	16	2	1	51	1
76	36	0	714	0	0	14	1	0	24	1
77	20	1	479	0	0	6	2	1	118	0
78	49	0	646	0	0	14	2	0	2	1
79	26	1	333	1	0	14	1	1	2	1
80	51	0	606	0	0	4	1	0	4	1
81	40	1	125	0	1	35	2	0	117	0
82	45	1	553	0	0	18	1	1	117	0
83	46	1	861	0	0	6	2	1	73	1
84	50	1	1249	0	0	11	1	1	3	1
85	55	0	214	0	0	3	2	1	114	0
86	46	0	582	1	1	42	0	1	1	1
87	32	0	697	0	1	22	0	1	1	1
88	36	0	148	0	0	0	1	1	1	1
89	20	1	603	0	1	34	0	0	17	1
90	24	1	881	0	0	59	1	1	79	1
91	50	0	320	1	0	2	1	1	61	1
92	39	1	580	0	1	45	2	0	22	1
93	19	0	451	0	1	20	2	0	1	1
94	33	1	691	1	1	24	2	1	65	1
95	39	1	88	0	0	6	0	1	2	1
96	53	0	157	0	0	2	2	1	26	1
97	41	0	385	0	1	22	2	1	96	1
98	43	1	370	0	0	5	1	1	106	0
99	39	0	708	0	0	0	1	1	105	0
100	48	0	1171	0	0	14	2	1	33	1
101	45	0	634	0	1	47	1	1	12	0
102	53	1	2069	0	0	14	3	1	18	1
103	37	0	399	0	0	3	2	1	102	0
104	35	1	508	1	0	0	1	1	3	1
105	56	1	506	0	0	5	1	1	102	0
106	56	0	283	0	0	9	1	0	102	0
107	33	0	1480	0	1	53	0	0	101	0
108	55	0	955	0	0	5	2	0	101	0
109	23	1	492	0	0	5	4	0	11	1
110	51	0	469	0	0	3	1	0	1	1
111	48	0	825	0	0	7	1	0	1	1

KIDNEY TRANSPLANT DATA

OBS	AGE	SEX	DIALY	DBT	PTX	BLOOD	MIS	ALG	MONTH	FAIL
112	57	1	1222	0	0	12	2	0	24	1
113	49	1	550	0	0	14	3	0	80	1
114	46	0	545	0	0	7	1	0	98	0
115	24	1	312	0	0	7	2	0	35	0
116	43	0	1270	0	0	7	2	0	20	1
117	19	0	615	0	0	17	2	0	2	0
118	53	0	65	0	0	9	2	0	89	1
119	40	1	271	0	1	11	1	0	96	0
120	19	0	128	0	0	7	1	1	96	0
121	27	0	57	0	0	10	3	0	1	1
122	21	1	629	0	0	9	2	0	95	0
123	37	1	1294	1	0	14	1	0	57	1
124	27	0	197	1	0	6	1	1	1	1
125	54	1	193	0	0	15	2	0	1	1
126	48	1	1581	0	0	12	1	0	54	1
127	42	0	452	0	1	11	1	0	30	1
128	22	1	387	0	0	17	2	0	93	0
129	31	1	311	1	0	21	2	0	6	1
130	25	1	1467	0	0	11	1	0	19	1
131	49	1	0	0	1	3	1	1	91	0
132	44	1	135	0	0	9	2	1	91	0
133	25	1	284	0	0	49	1	1	91	0
134	60	1	1373	0	0	15	2	1	72	1
135	61	1	1918	0	0	9	1	1	49	1
136	60	0	956	0	0	8	0	0	89	0
137	52	1	315	0	0	5	0	1	88	0
138	50	1	1802	0	0	6	0	1	88	0
139	47	0	1568	0	0	12	2	1	2	1
140	59	1	481	0	0	9	2	0	88	0
141	51	1	152	0	1	33	0	0	87	0
142	47	0	303	0	0	8	2	1	86	0
143	46	0	186	0	0	6	2	1	86	0
144	53	0	1267	0	2	8	2	1	2	1
145	38	1	672	1	0	2	0	1	85	0
146	36	1	140	1	0	2	1	0	85	0
147	40	0	1300	0	1	10	0	0	5	1
148	23	1	221	0	0	14	1	1	84	0
149	39	0	845	0	1	7	1	0	84	0
150	34	0	793	0	1	20	1	0	84	0
151	29	0	529	0	1	26	1	0	84	0
152	56	1	54	0	0	10	1	0	83	1
153	43	1	1701	0	0	6	3	0	23	1

KIDNEY TRANSPLANT DATA

OBS	AGE	SEX	DIALY	DBT	PTX	BLOOD	MIS	ALG	MONTH	FAIL
154	44	1	28	0	1	4	1	0	83	0
155	40	0	631	0	0	11	1	0	48	1
156	30	0	134	0	0	16	1	0	82	0
157	23	1	268	0	0	16	1	1	2	1
158	40	0	189	0	0	2	2	1	57	1
159	42	1	2194	0	1	33	2	0	3	1
160	56	0	333	0	0	17	2	1	81	0
161	55	1	394	0	0	25	1	1	80	0
162	48	1	339	0	0	16	1	1	79	0
163	52	0	667	0	0	12	1	0	79	0
164	40	0	340	0	0	4	1	1	45	1
165	32	1	342	0	0	3	3	0	1	1
166	40	0	357	1	0	5	2	0	79	0
167	49	0	375	0	0	19	2	1	56	1
168	37	0	121	0	0	4	2	1	79	0
169	35	0	223	0	0	8	2	1	5	1
170	36	0	147	0	0	5	1	1	77	0
171	32	1	284	1	1	34	2	1	77	0
172	39	1	274	0	0	12	2	1	77	0
173	56	1	99	0	0	7	2	1	77	0
174	63	1	1010	0	0	11	2	1	76	0
175	54	1	289	0	0	7	1	1	2	1
176	36	0	0	0	1	14	2	1	76	0
177	42	1	778	0	0	10	2	1	17	1
178	22	1	467	0	0	9	1	1	75	0
179	51	0	88	0	0	13	2	1	24	1
180	36	0	282	0	0	19	2	1	64	1
181	36	1	170	1	0	8	2	1	25	1
182	64	0	676	0	0	1	1	1	35	1
183	27	0	168	0	0	4	2	1	75	0
184	18	1	68	0	0	15	2	1	74	0
185	21	0	195	0	0	3	1	1	74	0
186	51	1	173	0	0	6	2	1	74	0
187	47	0	498	0	0	12	2	1	3	1
188	18	0	119	0	0	6	2	1	46	1
189	27	0	594	0	0	28	4	0	72	0
190	34	0	85	1	0	12	2	1	72	0
191	35	1	396	0	0	5	1	1	71	0
192	24	1	402	1	0	50	1	1	9	1
193	29	0	179	0	0	9	1	1	70	0
194	36	0	318	1	0	3	2	1	59	1
195	44	1	237	0	0	5	0	1	69	0

OBS	AGE	SEX	DIALY	DBT	PTX	BLOOD	MIS	ALG	MONTH	FAIL
196	38	1	545	0	0	8	2	1	1	1
197	24	1	577	0	0	6	1	1	69	0
198	52	0	540	0	0	21	2	1	69	0
199	35	0	1657	0	1	23	0	1	68	0
200	35	0	83	1	0	3	1	1	68	0
201	31	0	232	1	0	5	1	1	68	0
202	43	0	124	1	0	2	1	1	2	1
203	21	0	595	0	0	18	1	1	68	0
204	51	0	279	0	0	3	2	1	67	0
205	34	1	95	0	0	6	1	1	23	1
206	34	0	273	0	0	7	2	1	6	1
207	55	1	735	0	0	46	1	1	56	1
208	64	1	238	0	0	5	1	1	9	1
209	40	1	180	0	0	10	2	1	66	0
210	24	0	86	0	0	8	2	1	65	0
211	52	0	597	1	0	3	2	1	3	1
212	55	0	1163	1	0	12	2	1	4	1
213	40	0	1033	0	0	15	0	1	65	0
214	24	1	895	0	2	48	0	1	65	0
215	54	0	86	0	0	13	1	1	65	0
216	56	0	155	0	0	10	0	1	53	1
217	20	0	90	0	0	5	1	1	64	0
218	50	1	516	1	0	8	2	1	63	0
219	40	0	144	0	0	14	3	1	2	1
220	38	0	1095	1	1	9	0	1	3	1
221	52	0	443	0	0	5	0	1	62	0
222	21	0	202	1	0	14	0	1	62	0
223	68	0	1584	0	0	14	1	1	30	1
224	33	0	312	0	0	4	2	1	4	1
225	32	0	295	1	1	81	1	1	61	0
226	23	0	170	1	0	6	1	1	45	1
227	35	0	0	0	1	22	2	1	60	0
228	34	0	155	1	0	4	2	1	60	0
229	52	0	351	0	0	5	2	1	60	0
230	19	0	110	0	0	6	2	0	59	0
231	42	0	125	0	0	5	2	1	59	0
232	65	0	212	0	0	9	2	1	17	1
233	58	1	490	0	0	35	1	1	59	0
234	41	0	221	1	0	5	2	1	59	0
235	27	0	190	1	0	6	1	1	58	0
236	41	0	588	0	0	11	2	1	58	0
237	42	1	1330	0	0	18	1	1	57	0

KIDNEY TRANSPLANT DATA

OBS	AGE	SEX	DIALY	DBT	PTX	BLOOD	MIS	ALG	MONTH	FAIL
238	64	1	1051	0	0	42	1	1	38	1
239	67	0	1385	0	0	21	2	1	4	1
240	32	1	546	1	0	7	2	1	57	0
241	33	1	87	0	0	8	2	1	56	0
242	58	1	1568	0	0	18	2	1	28	1
243	34	0	575	1	0	9	2	1	55	0
244	29	0	122	0	0	4	2	1	55	0
245	49	0	176	1	0	6	2	1	9	1
246	30	0	938	0	0	8	2	0	54	0
247	49	0	265	0	0	5	1	1	2	1
248	65	0	548	0	0	9	2	1	5	1
249	42	1	187	1	0	8	2	1	4	1
250	36	0	189	1	0	5	2	1	10	1
251	37	1	402	0	0	5	2	1	53	0
252	19	1	124	0	0	5	1	1	53	0
253	32	1	231	0	0	8	2	1	8	1
254	33	0	205	0	0	5	2	1	52	0
255	28	0	122	0	0	5	1	1	52	0
256	42	0	194	0	0	16	1	1	1	1
257	25	0	126	1	0	6	0	1	52	0
258	24	1	159	1	0	5	0	1	52	0
259	30	0	292	0	0	5	1	1	52	0
260	29	0	284	1	0	8	2	1	51	0
261	67	0	531	0	0	12	2	1	51	0
262	48	1	280	0	0	8	1	0	11	1
263	58	0	118	1	0	16	1	1	37	1
264	31	1	578	0	1	7	0	1	42	1
265	30	0	170	0	0	12	2	1	49	0
266	25	0	600	0	0	5	1	1	49	0
267	25	0	108	1	0	5	1	1	49	0
268	38	0	0	0	1	45	2	1	49	0
269	48	0	1275	1	0	6	2	1	47	0
270	55	0	1587	0	0	6	2	1	47	0
271	62	1	420	0	0	4	1	1	47	0
272	56	1	256	0	0	9	2	1	2	1
273	17	1	180	0	0	9	2	1	46	0
274	36	1	153	0	1	10	2	1	46	0
275	50	1	249	0	0	7	1	1	46	0
276	37	0	432	1	0	11	0	1	46	0
277	40	0	0	0	1	13	1	1	45	0
278	35	0	428	0	0	9	2	1	45	0
279	62	1	620	0	0	10	1	1	32	1

KIDNEY TRANSPLANT DATA

OBS	AGE	SEX	DIALY	DBT	PTX	BLOOD	MIS	ALG	MONTH	FAIL
280	43	0	412	1	0	6	1	1	3	1
281	53	0	226	0	0	9	0	1	44	0
282	21	1	163	0	0	5	2	1	43	0
283	50	0	211	0	0	5	2	1	4	1
284	24	1	1675	0	1	15	1	1	43	0
285	39	0	278	0	0	9	2	1	20	1
286	39	0	189	0	0	5	2	1	43	0
287	23	0	324	0	0	7	2	1	43	0
288	39	0	136	1	0	6	2	1	3	1
289	29	0	197	1	0	5	1	1	42	0
290	37	1	0	0	1	16	2	1	42	0
291	32	0	71	0	0	5	2	1	42	0
292	61	1	451	0	0	6	3	1	42	0
293	34	0	365	0	0	5	1	1	42	0
294	24	1	253	1	0	4	2	1	42	0
295	27	0	111	0	0	7	1	0	1	1
296	35	0	41	0	1	18	2	0	41	0
297	45	0	629	0	0	9	2	1	40	0
298	23	0	560	0	0	12	1	1	19	1
299	25	1	781	0	0	6	2	1	2	1
300	30	1	211	1	0	6	1	1	40	0
301	26	1	336	1	0	5	1	1	40	0
302	31	0	271	0	0	7	2	1	40	0
303	35	0	487	1	0	8	2	1	39	0
304	33	1	201	1	0	5	1	1	38	0
305	50	1	218	1	0	17	1	1	37	0
306	35	0	160	0	0	5	2	1	37	0
307	28	0	252	0	1	21	2	1	37	0
308	50	0	144	0	0	5	2	1	37	0
309	27	1	240	0	0	15	2	0	1	1
310	56	0	300	0	0	9	0	1	4	1
311	47	1	453	0	0	11	2	1	30	1
312	40	0	489	0	0	21	1	1	36	0
313	40	0	646	0	0	9	2	1	36	0
314	40	0	283	0	0	7	2	1	35	0
315	44	0	341	1	0	9	2	1	35	0
316	45	0	287	0	0	6	2	1	35	0
317	39	0	161	0	0	8	2	1	35	0
318	44	1	629	0	0	7	2	1	34	0
319	32	0	1119	0	0	18	1	1	34	0
320	58	0	309	0	0	6	0	1	34	0
321	27	0	438	1	0	6	1	1	34	0

KIDNEY TRANSPLANT DATA

OBS	AGE	SEX	DIALY	DBT	PTX	BLOOD	MIS	ALG	MONTH	FAIL
322	24	1	109	0	0	5	1	1	34	0
323	47	0	860	1	0	17	3	1	34	0
324	31	1	1320	0	0	6	3	1	34	0
325	49	0	990	0	0	13	2	1	2	1
326	19	0	239	0	0	7	2	1	33	0
327	37	1	211	0	0	8	2	1	33	0
328	58	1	57	0	0	80	1	1	19	1
329	45	1	129	0	0	11	2	1	13	1
330	58	1	472	0	0	13	2	1	33	0
331	52	0	375	0	0	5	2	1	32	0
332	38	0	326	0	0	9	1	1	32	0
333	31	1	0	0	1	61	2	0	3	1
334	42	1	274	0	0	9	0	1	31	0
335	42	1	383	1	0	5	2	1	31	0
336	34	0	454	0	0	5	1	1	31	0
337	46	0	391	1	0	5	0	1	3	1
338	51	0	426	1	0	14	2	1	30	0
339	44	1	750	0	0	10	0	1	8	0
340	58	0	256	0	0	11	1	1	2	1
341	54	1	372	0	0	16	1	1	29	0
342	44	1	468	1	0	10	2	1	29	0
343	42	0	608	0	0	5	3	1	14	1
344	45	0	1540	0	0	14	2	1	27	0
345	21	0	628	0	0	8	2	1	28	0
346	33	1	655	1	0	6	1	1	10	1
347	65	0	243	0	0	44	2	1	7	1
348	48	0	1366	0	0	29	1	1	28	0
349	27	1	513	1	0	5	2	1	28	0
350	28	1	231	0	1	26	0	0	9	1
351	47	0	367	0	0	5	2	1	28	0
352	33	1	1479	1	0	6	1	1	2	1
353	58	0	260	0	0	7	1	1	15	1
354	53	1	1581	0	0	7	1	1	27	0
355	49	1	1515	1	0	5	0	1	14	1
356	22	0	217	0	0	5	2	1	26	0
357	60	0	454	0	0	5	2	1	26	0
358	63	0	146	0	0	5	2	1	25	0
359	18	1	280	0	0	5	0	1	1	1
360	19	1	178	0	0	5	1	1	24	0
361	34	1	193	0	0	13	1	1	24	1
362	21	1	477	0	0	14	0	1	24	0
363	41	1	140	1	1	8	1	1	23	0

KIDNEY TRANSPLANT DATA

OBS	AGE	SEX	DIALY	DBT	PTX	BLOOD	MIS	ALG	MONTH	FAIL
364	37	0	996	1	0	89	2	1	15	1
365	22	0	110	0	1	9	2	1	23	0
366	37	1	426	0	0	5	2	1	6	1
367	57	0	278	0	0	9	2	1	23	0
368	28	1	293	0	0	5	2	1	23	0
369	59	0	1151	0	0	11	3	1	5	1
370	41	1	1097	0	0	5	2	1	21	0
371	55	1	441	0	0	11	2	1	21	0
372	28	1	218	1	0	5	2	1	20	0
373	58	1	349	0	0	5	2	1	3	1
374	46	1	0	0	1	12	2	1	7	1
375	18	1	235	0	0	9	1	1	20	0
376	68	1	690	0	0	5	1	1	20	0
377	33	1	137	0	0	12	2	1	20	0
378	49	0	411	0	0	9	2	1	19	0
379	36	1	788	0	0	6	2	1	6	1
380	29	0	105	0	0	6	2	1	19	0
381	43	0	249	1	0	8	0	1	19	0
382	51	0	547	0	0	6	2	1	19	0
383	68	0	2007	0	0	25	3	1	6	1
384	41	0	243	1	0	6	1	1	16	1
385	33	0	287	0	0	6	2	1	1	1
386	43	0	493	0	0	7	1	1	19	0
387	18	1	155	0	0	9	0	1	19	0
388	54	0	265	1	0	16	2	1	17	0
389	31	0	246	0	0	10	2	1	17	0
390	46	0	547	0	0	9	2	1	17	0
391	33	1	534	1	0	7	2	1	17	0
392	29	0	719	1	0	9	2	1	16	0
393	18	0	447	0	0	10	1	1	16	0
394	21	0	160	0	0	8	2	1	16	0
395	56	0	464	0	0	31	2	1	15	1
396	41	1	859	1	2	42	0	1	15	0
397	51	1	773	0	0	10	2	1	15	0
398	29	0	62	0	1	18	2	1	15	0
399	27	1	714	0	0	49	1	1	15	0
400	58	1	671	0	0	7	1	1	14	0
401	23	1	434	0	0	17	2	1	14	0
402	25	0	957	1	0	10	0	1	14	0
403	28	0	599	0	0	13	0	1	14	0
404	52	0	1550	0	0	15	1	1	14	0
405	33	1	477	0	0	5	1	1	14	0

KIDNEY TRANSPLANT DATA

OBS	AGE	SEX	DIALY	DBT	PTX	BLOOD	MIS	ALG	MONTH	FAIL
406	35	1	1152	0	0	20	2	1	1	1
407	30	1	622	0	0	15	2	1	3	1
408	51	0	222	1	0	10	2	1	6	1
409	32	0	367	1	0	8	2	1	13	0
410	34	0	337	0	0	13	2	1	13	0
411	19	1	325	0	1	19	0	1	5	1
412	55	1	368	1	0	11	0	1	13	0
413	42	1	560	0	0	15	1	1	12	0
414	19	0	319	0	0	16	2	1	12	0
415	69	0	1467	0	2	33	1	1	1	1
416	47	1	2431	0	0	29	3	1	11	0
417	37	1	250	0	0	8	2	1	10	0
418	47	1	555	0	0	3	2	1	10	0
419	56	0	356	0	0	17	2	1	10	0
420	30	1	0	0	1	18	1	1	9	0
421	34	1	271	1	0	8	1	1	9	0
422	47	1	294	1	0	11	1	1	9	0
423	62	1	3463	0	0	3	2	1	8	0
424	51	1	472	1	0	14	2	1	8	0
425	40	0	226	0	0	15	1	1	3	1
426	80	0	388	0	0	13	2	1	8	0
427	72	0	1001	0	0	11	2	0	8	0
428	56	0	176	0	1	16	2	1	7	0
429	26	0	155	0	0	6	2	1	7	0
430	61	1	861	0	0	9	2	1	7	0
431	36	0	775	1	0	11	2	1	7	0
432	60	0	766	0	0	9	2	1	7	0
433	55	0	585	0	0	7	2	1	7	0
434	39	0	251	0	0	11	1	1	7	0
435	64	1	736	0	1	17	2	1	6	0
436	39	1	1136	0	1	21	0	1	6	0
437	24	1	239	1	0	7	2	1	6	0
438	32	0	1797	0	0	10	1	1	6	0
439	56	1	386	0	0	10	2	1	3	1
440	61	0	422	0	0	7	2	1	3	1
441	51	1	1640	0	1	36	3	1	6	0
442	33	1	315	1	0	10	1	1	6	0
443	42	0	169	0	1	24	2	1	6	0
444	37	0	130	1	0	8	2	1	1	1
445	50	1	457	1	0	4	1	1	6	0
446	38	1	342	0	0	6	3	1	6	0
447	60	0	339	0	0	27	1	1	5	0

OBS	AGE	SEX	DIALY	DBT	PTX	BLOOD	MIS	ALG	MONTH	FAIL
448	30	0	280	0	0	10	1	1	5	0
449	28	0	611	0	0	8	2	1	5	0
450	45	0	295	0	0	7	2	1	5	0
451	61	1	564	0	0	8	2	1	5	0
452	41	1	667	0	0	10	0	1	5	0
453	32	0	253	0	0	5	1	1	5	0
454	36	0	186	1	0	5	2	1	1	1
455	31	0	152	1	0	5	1	1	5	0
456	41	0	212	1	0	8	1	1	1	1
457	46	1	216	0	0	22	1	1	4	0
458	55	1	557	0	0	13	2	1	4	0
459	58	1	4156	0	0	20	1	1	4	0
460	18	1	649	0	0	11	2	1	3	0
461	32	0	186	1	0	7	2	1	3	0
462	68	1	1192	0	0	8	2	1	2	0
463	35	0	348	1	0	10	2	1	2	0
464	56	0	1322	0	0	9	0	1	2	0
465	50	0	1215	0	0	7	2	1	2	0
466	36	0	388	0	0	6	2	1	2	0
467	36	1	576	0	0	18	2	1	2	0
468	24	1	288	1	0	8	2	1	2	0
469	25	0	480	0	0	8	0	1	1	0

The eight (8) variables included are:
OBS: Observation/patient number
AGE: Age at transplant in years
SEX: 1 = female, 0 = male
DIALY: Duration of hemodialysis prior to transplant, in days
DBT: Diabetes; 1 = yes, 0 = no
PTX: Number of prior transplants
BLOOD: Amount of blood transfusion, in blood units
MIS: Mismatch score
ALG: Use of ALG, an immune suppression drug; 1 = yes, 0 = no
MONTH: Duration time starting from transplant, in months
FAIL: status of the new kidney; 1 = new kidney failed, 0 = functioning

APPENDIX B

Hemodialysis Data

OBS	MOS	AGE	SEX	STATUS	RACE	OBS	MOS	AGE	SEX	STATUS	RACE
1	76	64	1	1	1	259	2	29	1	0	1
2	89	71	0	1	1	260	61	67	1	0	1
3	10	24	0	0	1	261	7	23	0	0	1
4	64	71	0	1	1	262	34	62	0	0	1
5	6	42	0	1	2	263	19	26	1	0	1
6	119	59	1	1	1	264	12	49	1	0	1
7	70	52	1	1	1	265	3	23	0	0	1
8	124	42	0	0	1	266	80	41	0	0	1
9	57	60	0	1	1	267	10	59	1	1	1
10	0	22	1	0	1	268	15	21	0	0	1
11	16	50	0	0	1	269	14	63	0	0	1
12	64	71	1	1	1	270	24	53	0	0	1
13	16	58	1	0	1	271	9	23	1	0	1
14	122	60	0	0	1	272	11	31	0	0	1
15	6	53	1	0	1	273	41	21	0	0	1
16	27	40	0	0	1	274	7	71	0	1	1
17	120	23	1	0	1	275	55	63	1	0	1
18	28	38	0	0	1	276	2	27	0	0	1
19	74	70	1	0	1	277	2	18	1	0	1
20	22	50	1	0	1	278	57	28	0	0	1
21	19	54	0	0	1	279	56	39	1	0	1
22	93	64	0	1	1	280	20	19	0	0	1
23	119	54	1	0	2	281	37	64	1	1	1
24	9	31	1	0	1	282	18	23	0	0	2
25	4	19	1	0	1	283	14	47	1	0	1
26	24	51	0	1	1	284	18	37	0	0	1
27	9	62	0	0	2	285	7	40	1	0	1

HEMODIALYSIS DATA

OBS	MOS	AGE	SEX	STATUS	RACE	OBS	MOS	AGE	SEX	STATUS	RACE
28	34	38	1	0	1	286	4	25	1	0	1
29	6	28	1	1	1	287	3	71	1	0	1
30	3	71	1	1	1	288	4	30	1	0	1
31	118	67	0	0	1	289	9	54	0	0	1
32	19	53	1	1	1	290	7	34	1	0	1
33	10	43	0	1	1	291	79	64	0	1	1
34	114	62	1	1	1	292	22	65	0	0	1
35	22	62	1	0	1	293	104	66	0	0	1
36	90	71	0	1	1	294	87	49	1	0	1
37	17	64	0	0	1	295	80	63	1	1	1
38	10	46	1	0	1	296	14	35	1	0	1
39	48	81	1	1	1	297	7	29	1	0	2
40	115	52	0	0	1	298	13	35	0	0	1
41	9	57	0	1	1	299	9	21	0	0	1
42	3	26	0	0	1	300	102	56	1	0	1
43	5	49	0	0	1	301	23	73	1	1	1
44	2	28	1	0	2	302	4	38	1	0	1
45	19	81	0	1	1	303	4	30	0	0	1
46	2	42	0	0	1	304	32	74	0	1	1
47	6	46	1	0	1	305	21	48	0	0	1
48	59	71	0	1	1	306	16	60	1	1	1
49	100	57	0	0	1	307	13	64	1	0	1
50	5	36	1	0	1	308	18	53	0	0	1
51	3	54	0	0	1	309	1	66	0	0	1
52	75	25	0	0	1	310	22	25	1	0	1
53	72	35	0	0	1	311	53	78	0	1	1
54	2	35	0	0	1	312	5	30	1	0	1
55	6	21	1	0	1	313	2	68	0	1	2
56	4	33	0	1	1	314	86	33	0	0	1
57	6	27	1	0	2	315	8	40	0	0	1
58	52	49	0	0	2	316	90	56	0	0	1
59	11	65	0	0	1	317	12	52	1	0	1
60	61	72	0	1	1	318	37	29	1	0	1
61	54	30	0	0	1	319	40	66	1	0	1
62	40	62	0	1	1	320	61	48	0	1	1
63	7	65	1	1	1	321	42	42	1	0	1
64	44	81	0	1	1	322	14	67	0	1	1
65	45	17	1	0	1	323	66	73	1	1	1
66	38	50	1	0	1	324	89	63	0	0	1
67	8	44	0	0	1	325	40	34	0	0	1
68	46	61	0	0	1	326	29	43	1	0	1
69	44	29	1	0	2	327	5	42	1	0	1

HEMODIALYSIS DATA

OBS	MOS	AGE	SEX	STATUS	RACE	OBS	MOS	AGE	SEX	STATUS	RACE
70	37	64	0	0	1	328	20	24	1	0	1
71	40	57	1	1	2	329	3	17	1	0	1
72	1	27	1	0	1	330	80	39	1	0	1
73	8	18	1	0	1	331	3	25	1	0	1
74	65	73	1	1	1	332	1	70	0	0	1
75	43	28	0	0	1	333	87	59	0	0	1
76	95	67	0	0	1	334	5	41	1	0	1
77	2	18	0	0	1	335	86	22	1	0	2
78	9	51	1	0	1	336	86	57	0	0	1
79	5	32	1	0	1	337	3	20	1	0	1
80	83	84	1	1	1	338	86	65	1	0	1
81	6	29	0	0	1	339	14	44	1	0	1
82	9	34	1	0	2	340	19	22	0	0	1
83	27	18	0	0	1	341	13	48	1	0	1
84	16	57	0	0	1	342	29	37	0	0	1
85	79	77	1	1	1	343	33	47	1	0	1
86	62	59	1	1	1	344	8	81	0	0	1
87	17	40	1	0	2	345	25	23	0	0	1
88	11	21	1	1	2	346	4	29	1	0	1
89	2	16	0	0	1	347	84	25	0	0	2
90	8	64	0	0	1	348	84	56	1	0	1
91	21	10	1	0	1	349	51	48	1	0	1
92	11	52	1	0	1	350	20	61	0	0	1
93	12	43	1	0	1	351	67	68	1	1	1
94	3	69	1	0	1	352	4	42	1	0	1
95	15	55	0	0	1	353	8	50	0	0	1
96	8	66	1	1	1	354	6	27	0	0	1
97	7	21	0	0	1	355	25	42	0	0	1
98	76	75	0	1	1	356	49	14	0	0	1
99	3	19	1	0	1	357	5	76	1	0	1
100	27	25	1	0	1	358	72	60	1	0	1
101	50	44	0	0	2	359	13	33	1	0	1
102	41	36	1	1	1	360	75	76	0	0	1
103	9	49	1	0	1	361	22	21	1	0	1
104	3	27	1	0	1	362	20	81	1	1	1
105	16	26	1	0	1	363	45	77	1	1	1
106	37	42	1	1	1	364	6	17	0	0	1
107	80	54	0	0	1	365	9	39	1	0	1
108	10	29	1	0	1	366	11	10	1	0	1
109	18	25	0	0	1	367	42	25	1	0	1
110	14	61	0	0	1	368	8	16	1	0	1
111	6	45	1	0	1	369	16	39	1	0	1

OBS	MOS	AGE	SEX	STATUS	RACE	OBS	MOS	AGE	SEX	STATUS	RACE
112	56	65	0	1	1	370	3	72	0	1	1
113	8	30	0	0	1	371	70	44	1	0	1
114	9	28	0	0	1	372	21	32	0	0	1
115	79	31	1	0	1	373	9	27	1	0	2
116	4	30	1	0	1	374	65	76	0	1	1
117	3	33	0	0	1	375	6	53	1	0	1
118	76	68	0	0	1	376	19	82	0	0	1
119	7	33	1	0	1	377	63	50	1	0	1
120	14	35	1	0	2	378	44	28	1	0	2
121	18	22	1	0	1	379	33	81	0	1	1
122	12	59	0	1	1	380	40	46	0	0	1
123	77	29	1	0	2	381	69	71	0	0	1
124	66	35	0	0	1	382	18	19	0	0	2
125	25	42	1	0	1	383	34	66	0	0	1
126	18	79	1	1	1	384	6	39	1	0	1
127	36	39	0	0	1	385	66	52	1	0	2
128	21	77	1	0	2	386	9	31	1	0	1
129	4	20	0	0	1	387	25	23	1	0	1
130	21	39	1	0	1	388	14	33	1	0	1
131	7	31	0	0	2	389	11	35	1	0	2
132	54	38	0	0	1	390	18	21	1	0	2
133	75	78	1	0	1	391	4	73	0	1	1
134	3	55	0	0	1	392	5	21	0	0	1
135	15	61	0	0	1	393	1	32	0	0	1
136	4	28	1	0	1	394	59	76	0	1	1
137	21	43	0	0	1	395	19	73	0	1	1
138	74	69	0	0	1	396	51	41	1	0	1
139	11	30	1	0	2	397	65	76	0	0	1
140	8	56	0	0	1	398	4	83	1	1	1
141	23	56	0	0	1	399	30	63	1	1	1
142	65	33	1	0	1	400	18	67	1	1	1
143	50	60	0	0	1	401	21	22	0	0	1
144	65	55	0	0	1	402	39	50	0	0	1
145	15	34	1	0	1	403	3	24	0	0	1
146	19	56	1	0	1	404	13	49	1	0	2
147	2	30	1	0	1	405	57	68	0	0	1
148	4	66	1	0	1	406	46	36	0	0	1
149	55	34	0	0	1	407	23	25	0	0	1
150	43	53	1	0	1	408	5	60	0	0	1
151	16	35	1	0	1	409	22	56	0	0	1
152	5	57	1	0	1	410	56	67	1	0	1
153	12	28	0	0	1	411	59	35	1	0	1

HEMODIALYSIS DATA

OBS	MOS	AGE	SEX	STATUS	RACE	OBS	MOS	AGE	SEX	STATUS	RACE
154	9	39	1	0	1	412	7	66	1	0	1
155	9	45	1	0	1	413	5	55	1	0	1
156	6	46	1	0	1	414	5	58	0	0	1
157	2	32	1	0	1	415	40	18	1	1	1
158	12	52	1	0	1	416	17	70	0	1	1
159	19	54	1	0	1	417	4	54	1	0	1
160	14	77	1	1	1	418	39	72	0	1	1
161	64	67	0	0	1	419	35	74	0	1	1
162	31	59	0	0	1	420	32	48	1	0	2
163	1	45	1	0	1	421	20	28	1	0	1
164	62	68	0	0	1	422	1	52	1	0	1
165	11	27	1	0	1	423	45	18	0	0	1
166	60	61	0	1	1	424	53	27	1	0	2
167	5	35	1	0	1	425	32	66	0	0	1
168	9	66	0	1	1	426	2	24	1	0	1
169	11	50	1	0	1	427	52	34	0	0	1
170	11	38	1	0	1	428	13	49	1	0	1
171	3	51	0	0	1	429	8	42	0	0	2
172	61	39	0	0	2	430	27	23	0	0	1
173	61	68	0	0	1	431	5	23	1	0	1
174	58	67	0	0	1	432	50	56	0	0	1
175	5	50	1	0	1	433	18	52	1	0	1
176	8	9	1	0	1	434	21	60	1	1	1
177	12	67	0	0	1	435	3	27	0	0	1
178	12	54	0	0	1	436	22	61	0	0	1
179	14	58	0	1	1	437	6	19	0	0	1
180	2	41	1	0	1	438	49	28	0	0	1
181	10	17	1	0	1	439	22	39	0	0	1
182	15	66	0	1	1	440	16	32	0	0	1
183	23	50	0	0	1	441	51	71	0	0	1
184	15	59	1	0	1	442	48	60	0	0	1
185	29	67	1	0	1	443	18	46	0	0	1
186	4	27	1	0	1	444	17	36	0	0	1
187	16	20	0	0	1	445	45	33	1	0	1
188	12	70	0	1	1	446	11	80	1	0	1
189	47	35	0	0	2	447	7	73	0	0	1
190	28	37	1	0	2	448	13	35	1	0	2
191	46	52	1	0	1	449	8	31	1	0	1
192	37	22	1	0	2	450	31	50	0	0	1
193	4	76	0	1	1	451	5	26	1	0	1
194	34	18	0	0	2	452	31	68	0	0	1
195	45	60	1	0	1	453	31	75	0	0	1

OBS	MOS	AGE	SEX	STATUS	RACE	OBS	MOS	AGE	SEX	STATUS	RACE
196	32	34	0	0	1	454	6	43	1	0	1
197	30	72	0	1	1	455	30	37	1	0	1
198	7	33	0	0	2	456	8	63	1	1	1
199	21	16	0	0	2	457	29	76	1	1	1
200	30	53	0	0	1	458	25	35	0	0	1
201	4	34	0	0	1	459	29	67	0	0	1
202	11	35	1	0	1	460	2	25	0	0	1
203	3	30	1	0	1	461	29	67	1	0	1
204	2	27	1	0	1	462	28	62	0	0	1
205	10	19	1	0	1	463	0	31	0	0	1
206	37	64	0	0	1	464	12	26	1	0	1
207	31	50	0	1	2	465	1	38	0	0	2
208	15	59	0	0	1	466	10	59	0	0	1
209	5	19	0	0	1	467	28	73	0	0	1
210	13	80	1	0	1	468	2	55	0	0	1
211	18	35	0	0	1	469	26	56	0	0	1
212	8	58	0	0	1	470	4	81	0	0	1
213	16	77	1	1	1	471	26	63	0	0	1
214	39	78	0	0	1	472	15	32	0	0	1
215	8	56	1	0	1	473	26	66	0	0	1
216	24	74	1	1	1	474	25	61	1	0	2
217	38	75	1	0	1	475	24	21	0	0	1
218	5	29	1	0	2	476	15	40	1	0	1
219	12	18	1	0	1	477	24	23	1	0	2
220	38	80	0	0	1	478	1	78	0	0	1
221	8	37	0	0	1	479	23	32	0	0	1
222	3	51	0	0	1	480	2	69	0	0	1
223	11	37	0	0	1	481	7	44	0	0	2
224	23	52	1	0	1	482	20	61	1	0	1
225	30	25	1	0	1	483	22	65	0	0	1
226	35	46	1	0	1	484	4	57	0	0	1
227	3	47	1	0	2	485	22	33	1	0	2
228	35	77	1	0	1	486	21	60	1	0	1
229	28	41	0	0	1	487	20	26	0	0	1
230	34	45	1	0	1	488	20	45	1	0	1
231	29	32	1	0	1	489	7	44	1	0	1
232	31	53	1	0	1	490	20	70	1	0	1
233	17	59	1	0	1	491	19	36	1	0	1
234	6	21	1	0	1	492	2	32	0	0	1
235	18	40	0	0	1	493	7	56	1	0	2
236	19	43	0	0	1	494	3	15	1	0	1
237	18	47	1	0	2	495	14	27	1	0	2

OBS	MOS	AGE	SEX	STATUS	RACE	OBS	MOS	AGE	SEX	STATUS	RACE
238	12	25	0	0	1	496	6	61	1	0	2
239	18	43	0	0	2	497	14	33	0	0	1
240	13	23	1	0	1	498	14	39	1	0	1
241	16	55	1	0	1	499	2	78	1	0	1
242	9	40	0	0	1	500	12	62	0	0	1
243	17	17	0	0	1	501	12	68	1	0	1
244	16	76	0	0	1	502	12	63	1	0	1
245	6	56	1	0	1	503	11	56	0	0	2
246	12	25	1	0	1	504	11	27	1	0	1
247	12	25	0	0	1	505	10	64	0	0	1
248	16	65	0	0	1	506	11	76	0	0	1
249	16	30	0	0	1	507	10	60	0	0	1
250	16	54	0	0	2	508	10	51	1	0	1
251	9	90	1	1	1	509	3	85	1	1	1
252	10	32	1	0	1	510	9	47	0	0	1
253	11	31	0	0	1	511	2	23	1	0	1
254	14	63	0	0	1	512	8	84	0	0	1
255	2	47	0	0	1	513	1	85	0	1	1
256	2	69	1	0	1	514	7	63	1	0	1
257	14	41	1	0	2	515	8	55	1	0	1
258	4	56	0	0	2	516	7	79	0	0	1

The five (5) variables included are:
OBS: Observation/patient number
MOS: Duration of dialysis in months
AGE: Age at the start of dialysis in years
SEX: 0 = female, 1 = male
STATUS: 0 = alive, 1 = dead
RACE: 1 = white, 2 = others

APPENDIX C

Ventilation Tube Data

CHILD	TRT	EAR	TIME	ST	CHILD	TRT	EAR	TIME	ST
01	1	1	3.10	1	40	2	1	15.40	1
		2	4.80	0			2	9.20	1
02	1	1	12.70	1	41	2	1	9.30	1
		2	6.00	1			2	9.30	1
03	1	1	3.10	1	42	2	1	15.00	1
		2	6.40	1			2	.90	1
04	1	1	8.50	1	43	2	1	15.00	1
		2	12.70	1			2	11.90	1
05	1	1	9.10	1	44	2	1	17.80	1
		2	9.10	1			2	12.20	1
06	1	1	.50	1	45	2	1	5.90	1
		2	5.10	1			2	8.70	1
07	1	1	18.10	1	46	2	1	8.90	1
		2	15.00	1			2	12.60	1
08	1	1	6.00	1	47	2	1	.60	1
		2	6.00	1			2	5.70	1
09	1	1	6.40	1	48	2	1	6.00	1
		2	6.40	1			2	9.40	1
10	1	1	4.40	0	49	2	1	14.60	1
		2	1.30	1			2	9.00	1
11	1	1	12.80	1	50	2	1	12.10	1
		2	12.80	1			2	2.90	1
12	1	1	8.80	1	51	2	1	3.00	1
		2	8.80	1			2	3.00	1
13	1	1	2.80	0	52	2	1	24.90	1
		2	2.80	0			2	8.70	1

VENTILATION TUBE DATA

CHILD	TRT	EAR	TIME	ST	CHILD	TRT	EAR	TIME	ST
14	1	1	9.30	1	53	2	1	5.20	1
		2	27.10	1			2	9.00	1
15	1	1	6.10	1	54	2	1	24.30	1
		2	6.10	1			2	18.80	1
16	1	1	17.90	1	55	2	1	15.20	1
		2	20.90	1			2	12.50	1
17	1	1	9.30	1	56	2	1	33.00	1
		2	3.10	1			2	12.10	1
18	1	1	2.90	1	57	2	1	13.10	1
		2	1.00	1			2	.70	1
19	1	1	9.10	1	58	2	1	6.10	1
		2	9.10	1			2	17.10	0
20	1	1	5.80	1	59	2	1	9.50	1
		2	9.30	1			2	3.40	1
21	1	1	2.90	1	60	2	1	15.10	1
		2	1.10	1			2	17.80	1
22	1	1	6.20	1	61	2	1	5.80	1
		2	9.20	1			2	5.80	1
23	1	1	1.10	0	62	2	1	.60	1
		2	1.10	0			2	3.00	1
24	1	1	6.00	1	63	2	1	2.80	1
		2	10.70	1			2	1.60	1
25	1	1	6.20	1	64	2	1	6.20	1
		2	6.20	1			2	9.00	1
26	1	1	9.30	1	65	2	1	8.70	1
		2	19.30	0			2	3.40	1
27	1	1	.80	1	66	2	1	20.90	0
		2	.80	1			2	3.40	1
28	1	1	13.30	1	67	2	1	9.20	1
		2	13.30	1			2	6.00	1
29	1	1	.80	1	68	2	1	6.40	1
		2	8.10	1			2	14.30	0
30	1	1	3.00	1	69	2	1	8.80	1
		2	15.80	1			2	8.80	1
31	1	1	9.40	1	70	2	1	18.50	1
		2	9.40	1			2	13.30	1
32	1	1	3.10	1	71	2	1	12.20	1
		2	3.10	1			2	12.20	1
33	1	1	7.60	1	72	2	1	12.50	0
		2	10.10	1			2	8.80	1
34	1	1	5.50	1	73	2	1	8.50	1
		2	5.50	1			2	21.70	1

VENTILATION TUBE DATA

CHILD	TRT	EAR	TIME	ST	CHILD	TRT	EAR	TIME	ST
35	1	1	.70	1	74	2	1	1.80	1
		2	.70	1			2	20.70	1
36	1	1	7.00	1	75	2	1	6.20	1
		2	7.00	1			2	9.00	1
37	1	1	11.70	1	76	2	1	9.70	1
		2	3.10	1			2	11.10	0
38	1	1	14.30	1	77	2	1	6.00	1
		2	3.20	1			2	6.00	1
39	2	1	11.90	1	78	2	1	8.70	1
		2	8.80	1			2	8.70	1

TRT: Treatment; 1 = control, 2 = medical
EAR: 1 = right, 2 = left
TIME: tube duration, in months
ST: tube functioning status; 1 = failed, 2 = censored

References

Ahlquist, D. A., D. B. McGill, S. Schwartz, and W. F. Taylor (1985). Fecal blood levels in health and disease: A study using HemoQuant. *New England Journal of Medicine* 314: 1422.

Bamber, D. (1975). The area above the ordinal dominance graph and the area below the receiver operating graph. *Journal of Mathematical Psychology* 12: 387–415.

Board, J. W. (1949). Maximum likelihood estimates of proportion of patients cured by cancer therapy. *Journal of the Royal Statistical Society* B11: 15–44.

Breslow, N. (1970). A generalized Kruskal-Wallis test for comparing K samples subject to unequal patterns of censorship. *Biometrika* 57: 579–594.

Breslow, N. (1974). Covariance analysis of censored survival data. *Biometrics* 30: 89–99.

Breslow, N. (1975). Analysis of survival data under the proportional hazards model. *International Statistics Review* 30: 43–54.

Breslow, N. (1982). Covariance adjustment of relative-risk estimates in matched studies. *Biometrics* 38: 661–672.

Breslow, N. and N. E. Day (1980). *Statistical Methods in Cancer Research I; The Analysis of Case-Control Studies*. Lyon, France: International Agency for Research in Cancer.

Cohen, A. C. (1965). Maximum likelihood estimation in the Weibull distribution based on complete and censored samples. *Technometrics* 7: 579–588.

Conover, W. J. and R. L. Iman (1981). Rank transformation as a bridge between parametric and nonparametric statistics. *American Statistician* 35: 124–129.

Cox, D. R. (1953). Some simple approximate tests for Poisson variates. *Biometrika* 40: 354–360.

Cox, D. R. (1972). Regression models and life tables. *Journal of the Royal Statistical Society* B34: 187–220.

Cox, D. R. (1975). Partial likelihood. *Biometrika* 62: 269–276.

Cox, D. R. (1979). A note on graphical analysis of survival data. *Biometrika* 66: 188–190.

Cox, D. R. and D. Oakes (1984). *Analysis of Survival Data*. London: Chapman and Hall.

Crowley, J. and M. Hu (1977). Covariance analysis of heart transplant survival data. *Journal of the American Statistical Association* 72: 27–36.

Cuzick, J. (1985). A Wilcoxon-type test for trend. *Statistics in Medicine* 4: 87–90.

Efron, B. (1977). The efficiency of Cox's likelihood function for censored data. *Journal of the American Statistical Association* 72: 557–565.

Feigl, P. and M. Zelen (1965). Estimation of exponential survival probabilities with concomitant information. *Biometrics* 21: 826–838.

Fleming, T. R., J. R. O'Fallon, P. C. O'Brien and D. P. Harrington (1980). Modified Kolmogorov-Smirnov test procedures with application to arbitrarily right-censored data. *Biometrics* 36: 607–625.

Fox, A. J. and P. F. Collier (1976). Low mortality rates in industrial cohort studies due to selection for work and survival in the industry. *British Journal of Preventive and Social Medicine* 30: 225–230.

Freireich, J., E. Gehan, E. Frei, et al. (1963). The effect of 6-mercaptoburine on the doration of steroid-induced remissions in acute leukemia: a model for evaluation of other potentially useful therapy. *Blood* 21: 699–716.

Gail, M. H., T. J. Santner, and C. C. Brown (1980). An analysis of comparative carcinogenesis experiments based on multiple times to tumor. *Biometrics* 36: 255–266.

Gehan, E. A. (1965a). A generalized Wilcoxon test for comparing arbitrarily singly-censored samples. *Biometrika* 52: 203–223.

Gehan, E. A. (1965b). A generalized two-sample Wilcoxon test for doubly censored data. *Biometrika* 52: 650–653.

Glasser, M. (1967). Exponential survival with covariance. *Journal of the American Statistical Association* 62: 561–568.

Goldman, A. I. (1984). Survivorship analysis when cure is a possibility: A Monte Carlo Study. *Statistics in Medicine* 3: 153–163.

Goldman, A. I. (1992). Eventcharts: visualizing survival and other timed-events data. *The American Statistician* 46: 13–18.

Greenwood, M. (1926). The natural duration of cancer. *Reports on Public Health and Medical Subjects*. London: Her Majesty's Stationary Office, Vol. 33, pp. 1–26.

Hanley, J. A. and B. J. McNeil (1982). The meaning and use of the area under a receiver operating characteristic (ROC) curve. *Radiology* 143: 29–36.

REFERENCES

Herbst, A. L., H. Ulfelder, and D. C. Poskanzer (1971). Adenocarcinoma of the vagina. *New England Journal of Medicine* 284: 878–881.

Hlatky, M. A., D. B. Pryor, F. E. Harrell, R. M. Califf, D. B. Mark, and R. A. Rosati (1984). Factors affecting sensitivity and specificity of exercise electrocardiography—Multivariate analysis. *American Journal of Medicine* 77: 64–71.

Hoeffding, W. (1954). On the distribution of expected values of order statistics. *Annals of Mathematical Statistics* 24: 93–100.

Holt, J.D. and R.L. Prentice (1974). Survival analyses in twin studies and matched pair experiments. *Biometrika* 61: 17–30.

Huang, Y. (1994). A class of tests against stochastically ordered alternatives for censored survival data. Masters Plan B Project. Division of Biostatistics, University of Minnesota.

Johnson, N. L. and S. Kotz (1969). *Discrete Distributions*. New York: Wiley.

Jones, M. P., T. W. O'Gorman, J. H. Lemke, and R. F. Woolson (1989). A Monte Carlo investigation of homogeneity tests of the odds ratio under various sample size configurations. *Biometrics* 45: 171–181.

Jonckheere, A. R. (1954). A distribution-free k-sample test against ordered alternatives. *Biometrika* 41: 133–145.

Kadane, J. B. (1971). A moment problem for order statistics. *Annals of Mathematical Statistics* 42: 745–751.

Kalbfleisch, J. D. and R. L. Prentice (1973). Marginal likelihoods based on Cox's regression and life table model. *Biometrika* 60: 267–278.

Kalbfleisch, J. D. and R. L. Prentice (1980). *The Statistical Analysis of Failure Time Data*. New York: Wiley.

Kaplan, E. L. and P. Meier (1958). Nonparametric estimation from incomplete observations. *Journal of the American Statistical Association* 53: 457–481.

King, M., D. M. Bailey, D. G. Gibson, J. V. Pitha, and P. B. McCay (1979). Incidence and growth of mammary tumors induced by 7,12-dimethylbenz(α)antheacene as related to the dietary content of fat and antioxidant. *Journal of the National Cancer Institute* 63: 656–664.

Koehler, K. J. and P. G. McGovern (1990). An application of the LFP survival model to the smoking cessation data. *Statistics in Medicine* 9: 409–421.

Le, C. T. (1988a). Testing for linear trends using correlated otolaryngology or ophthalmology data. *Biometrics* 44: 299–303.

Le, C. T. (1988b). A new rank test against ordered alternatives in k-sample problems. *Biometrical Journal* 30: 87–92.

Le, C. T. (1990). A different approach to the analysis of person-years data. *Biometrical Journal* 32: 615–619.

Le, C. T. (1991). The relationship between two statistical tests for the comparison of two expontial distributions. *Biometrical Journal* 33: 369–372.

Le, C. T., K. A. Daly, R. H. Mongolis, B. R. Lindgren, and G. S. Giebink. A clinical profile of otitis media. *Archives of Otolaryngology* 118: 1225–1228.

Le, C. T. (1993). A test for linear trend in constants hazards and its application to a problem in occupational health. *Biometrics* 49: 1220–1224.

Le, C. T. and D. Zelterman (1992). Tests of goodness-of-fit for Cox's proportional hazards models. *Biometrical Journal* 34: 557–566.

Le, C. T. and P. M. Grambsch (1994). Tests of association between survival time and a continuous covariate. *Communications in Statistics—Theory and Methods* 23: 1009–1019.

Le, C. T., B. R. Lindgren, D. Zelterman, and A. J. Umen (1991a). Homogeneity of the relative risk in cohort studies. *Statistics in Medicine* 10: 1267–1272.

Le, C. T., B. R. Lindgren, D. Zelterman, and A. J. Umen (1991b). Homogeneity of the relative risk in cohort studies. *Statistics in Medicine* 10: 1267–1272.

Le, C. T., P. M. Grambsch, and T. A. Louis (1994). Association between survival time and ordinal covariates. *Biometrics* 50: 213–219.

Le, C. T., P. M. Grambsch, D. Zelterman, and B. R. Lindgren (1993). Evaluation of staging systems for colorectal cancer. *Biometrical Journal* 35: 701–705.

Le, C. T. and B. R. Lindgren (1996). Duration of ventilating tubes: A test for comparing two clustered samples of censored data. *Biometrics* 52: 328–334.

Lee, E. I. (1980). *Statistical Methods for Survival Data Analysis.* Belmont, CA: Lifetime Learning Publications.

Levene, H. (1960). Robust tests for equality of variances. In: *Contributions to Probability and Statistics*, E. I. Olkin (Ed.) Palo Alto, CA: Stanford University Press, 278–292.

Liang, K. Y. and S. G. Self (1985). Tests for homogeneity of odds ratio when the data are sparse. *Biometrika* 72: 353–358.

Mack, T. M., M. C. Pike, B. E. Henderson, et al. (1976). Estrogens and endometrial cancer in a retirement community. *New England Journal of Medicine* 294: 1262–1267.

Mantel, N. (1967). Ranking procedures for arbitrarily restricted observations. *Biometrics* 23: 65–78.

Mantel, N. and W. Haenszel (1959). Statistical aspects of the analysis of data from retrospective studies of disease. *Journal of the National Cancer Institute* 22: 719–748.

Multiple Risk Factor Intervention Trial Research Group (1982). Multiple risk factor intervention trial: risk factor changes and mortality results. *Journal of the American Medical Association* 248: 1465–1477.

Nagelkerke, N. J. D., J. Oosting, and A. A. M. Hart (1984). A simple test for goodness of fit of Cox's proportional hazards model. *Biometrics* 40: 483–486.

Nelson, W. (1972). Theory and applications of hazard plotting for censored failure data. *Technometrics* 14: 945–966.

Newhouse, M. L. and G. Berry (1979). Patterns of mortality in asbestos factory workers in London. *Annals of the New York Academy of Sciences* 330: 53–60.

O'Brien, P. C. (1978). A nonparametric test for association with censored data. *Biometrics* 34: 243–250.

O'Brien, P. C. and T. R. Fleming (1987). A paired Prentice-Wilcoxon test for censored paired data. *Biometrics* 43: 169–180.

O'Quigley, J. and R. L. Prentice (1991). Nonparametric tests of association between survival time and continuously measured covariates: The logit-rank and associated procedures. *Biometrics* 47: 117–127.

Oakes, D. (1982). A concordance test for independence in the presence of censoring. *Biomtrics* 38: 451–455.

Peto, R. and J. Peto (1972). Asymptotically efficient rank invariant procedures. *Journal of the Royal Statistical Society* A135: 185–207.

Rosner, B. (1982). Statistical methods in ophthalmology: An adjustment for the intra-class correlation between eyes. *Biometrics* 38: 105–114.

Schoenfeld, D. (1980). Goodness of fit tests for the proportional hazards regression model. *Biometrika* 67: 145–154.

Sukhatme, S. and Beam, C. A. (1994). Stratification in nonparametric ROC studies. *Biometrics* 50: 149–163.

Tanner, M. A. and W. H. Wong (1984). Data-based nonparametric estimation of the hazard function and applications to model diagnostics and explanatory analysis. *Journal of the American Statistical Association* 79: 174–182.

Tarone, R. E. (1975). Tests for trend in life table analysis. *Biometrika* 62: 679–682.

Tarone, R. E. and J. Ware (1977). On distribution-free tests for equality of survival distributions. *Biometrika* 64: 156–160.

Therneau, T. M., P. M. Grambsch, and T. R. Fleming (1990). Martingale-based residuals and survival models. *Biometrika* 77: 147–160.

Tsiatis, A. A. (1981). A large sample study of Cox's regression model. *Annals of Statistics* 9: 93–108.

Ury, K.K. and A. D. Wiggins (1985) (Letter to the Editor). Another shortcut method for calculating the confidence interval of a Poisson variable (or of a Standardized Mortality Ratio). *American Journal of Epidemiology* 122: 197–198.

Wei, L. J., D. Y. Lin, and D. Weissfeld (1989). Regression analysis of multivariate incomplete failure time data by modeling marginal distributions. *Journal of the American Statistical Association* 84: 1065–1073.

Weier, D. R. and A. P. Basu (1980). An investigation of Kendall's tau modified for censored data with applications. *Journal of Statistical Planning and Inference* 4: 381–390.

Wilcoxon, F. (1945). Individual comparison by ranking methods. *Biometrics* 1: 80–83.

Wolfe, R. A., D. S. Gaylin, F. K. Port, P. J. Held, and C. L. Wood (1992). Using USRDS generated mortality tables to compare local ESRD mortality rates to national rates. *Kidney International* 42: 991–996.

Zelterman, D. (1992). A statistical distribution with an unbounded hazard function and its application to a theory in demography. *Biometrics* 48: 807–818.

Zelterman, D. and C. T. Le (1991). Tests of homogeneity for the relative risk in multiply-matched case-control studies. *Biometrics* 47: 751–756.

Zelterman, D., P. M. Grambsch, C. T. Le, Z. Ma, and J. W. Curtsinger (1994). Piecewise exponential survival curves with smooth transitions. *Mathematical Biosciences* 120: 233–250.

Zelterman, D., C. T. Le, and T. A. Louis (1995). Bootstrap techniques for proportional hazards models with censored observations. *Statistics and Computing* (in press).

Index

actuarial method 57
adjusted survival rate 64
area under the curve 6

backward elimination 188
baseline survival function 192
bathtub curve 8
BMDP package 43

case-control study 2
censoring 35
clinical trial 3
clustered data 148
coefficient of correlation 32
cohort study 2, 77
comparison of two groups 99
   nonparametric methods 99
   exponential samples 114
comparison of several groups 107
computational implementation 43
concordance 30
constant hazard 9, 142
contingency table 31
correlation analysis 162
covariate 13, 36, 162
Cox F test 118
Cox–Mantel test 101
crossing-curves alternative 106, 109
cumulative distribution function 8
cumulative hazard function 9, 86
cure model 105
current life table 66
Cuzick test 25

data 2
deviance residual 204
discordance 30
discrete model 11, 54
drop out 35
duration time 36

effect modification 180
empirical cumulative distribution
   function 17
ending event 4
enrollment period 3
epidemiologic matched study 209
estimation 51
   cumulative hazard function 86
   exponential model 70
   Kendall tau 87
   maximum likelihood 69
   parameters 60
   parameters in survival models 67
   survival function 52
   Weibull model 72
event chart 39
expected mortality 79
expected number of deaths 43
exponential model 9, 34, 68, 70, 114

follow-up period 3
force of mortality 8
forward selection 188

Gehan–Breslow test 101
Gompertz model 68
Greenwood's formula 55

hazard function 4, 8, 9, 10
healthy worker effect 84, 143
homogeneity of odds ratio 214

incomplete data 35

Jonckheere test 25

Kaplan–Meier 53
Kendall 30
Kolmogorov statistic 17
Kolmogorov–Smirnov statistic 18
Kruskal–Wallis test 23

Le test 26
Le–Grambsch–Louis test 138
Levene test 110
life span 4
likelihood function 33, 54, 69
likelihood ratio test 187
linearity 179
log-normal model 68
log-rank test 101
loss to follow-up 35

Mann–Whitney test 21
Mantel–Haenszel odds ratio 213
martingale residual 203
maximum-likelihood estimation 33
maximum observable time 41
mean 4, 5, 60
mean residual life 7
measure of association 165
measurement scale 3, 167
median 4, 5, 18, 60
multiple regression 176

nonparametric method 29, 98, 173
number of deaths 41

ordered alternative 133

pair-matched data 112
parameter 51, 60, 67
person-years data 125
Peto likelihood function 165
piecewise exponential model 128

polynomial regression 181
potential censoring time 69
power 121
Prentice–Wilcoxon test 112
probability density function 6
probability mass 10
product limit method 53
prognostic factor 162
proportional hazards model 12, 80, 102, 151, 163, 177, 217
prospective study 2

random censoring 35
rank correlation 28, 87
rate 8
regression analysis 12, 162
  multiple regression 176
  simple regression 163
  stepwise regression 188
relative risk 12
residual analysis 203
retrospective design 2
risk factor 162
risk function 8
risk set 102
ROC curve 216
ROC function 216

sample size determination 121
SAS program 56, 58, 106, 169, 172, 179, 192, 199, 202, 205, 212
score test 115, 146
shape parameter 10
Siegel–Tukey test 22
simple regression 163
Spearman 29
standard population 79
standardization of rate 62
standardized mortality ratio 74, 81, 143
starting point 3
step function 11
stepwise regression 187
stratification 207
study design 2
study termination 36
survival curve 4, 56
survival function 4, 52
survival model 68

survival rate 53, 59
survival status 36
survival time 3

Tarone test 137
Taronr–Ware class of tests 98
test of association 169
test of goodness of fit 197, 202
testing hypotheses 182
time-dependent covariate 196
time origin 4
trend in survival 132

uniform entry 41
Ury and Wiggins method 82

variable 2

Weibull model 10, 68, 72
Wilcoxon test 20, 101, 104

WILEY SERIES IN PROBABILITY AND STATISTICS

ESTABLISHED BY WALTER A. SHEWHART AND SAMUEL S. WILKS

Editors
Vic Barnett, Ralph A. Bradley, Noel A. C. Cressie, Nicholas I. Fisher, Iain M. Johnstone, J. B. Kadane, David G. Kendall, David W. Scott, Bernard W. Silverman, Adrian F. M. Smith, Jozef L. Teugels, Geoffrey S. Watson; J. Stuart Hunter, Emeritus

Probability and Statistics
ANDERSON · An Introduction to Multivariate Statistical Analysis, *Second Edition*
*ANDERSON · The Statistical Analysis of Time Series
ARNOLD, BALAKRISHNAN, and NAGARAJA · A First Course in Order Statistics
BACCELLI, COHEN, OLSDER, and QUADRAT · Synchronization and Linearity: An Algebra for Discrete Event Systems
BARTOSZYNSKI and NIEWIADOMSKA-BUGAJ · Probability and Statistical Inference
BERNARDO and SMITH · Bayesian Statistical Concepts and Theory
BHATTACHARYYA and JOHNSON · Statistical Concepts and Methods
BILLINGSLEY · Convergence of Probability Measures
BILLINGSLEY · Probability and Measure, *Second Edition*
BOROVKOV · Asymptotic Methods in Queuing Theory
BRANDT, FRANKEN, and LISEK · Stationary Stochastic Models
CAINES · Linear Stochastic Systems
CAIROLI and DALANG · Sequential Stochastic Optimization
CHEN · Recursive Estimation and Control for Stochastic Systems
CONSTANTINE · Combinatorial Theory and Statistical Design
COOK and WEISBERG · An Introduction to Regression Graphics
COVER and THOMAS · Elements of Information Theory
CSÖRGŐ and HORVÁTH · Weighted Approximations in Probability Statistics
*DOOB · Stochastic Processes
DUDEWICZ and MISHRA · Modern Mathematical Statistics
DUPUIS and ELLIS · A Weak Convergence Approach to the Theory of Large Deviations
ETHIER and KURTZ · Markov Processes: Characterization and Convergence
FELLER · An Introduction to Probability Theory and Its Applications, Volume 1, *Third Edition*, Revised; Volume II, *Second Edition*
FREEMAN and SMITH · Aspects of Uncertainty: A Tribute to D. V. Lindley
FULLER · Introduction to Statistical Time Series, *Second Edition*
FULLER · Measurement Error Models
GHOSH, MUKHOPADHYAY, and SEN · Sequential Estimation
GIFI · Nonlinear Multivariate Analysis
GUTTORP · Statistical Inference for Branching Processes
HALD · A History of Probability and Statistics and Their Applications before 1750
HALL · Introduction to the Theory of Coverage Processes
HANNAN and DEISTLER · The Statistical Theory of Linear Systems
HEDAYAT and SINHA · Design and Inference in Finite Population Sampling
HOEL · Introduction to Mathematical Statistics, *Fifth Edition*
HUBER · Robust Statistics
IMAN and CONOVER · A Modern Approach to Statistics
JOHNSON and KOTZ · Leading Personalities in Statistical Sciences: From the Seventeenth Century to the Present

*Now available in a lower priced paperback edition in the Wiley Classics Library.

Probability and Statistics (Continued)

JUREK and MASON · Operator-Limit Distributions in Probability Theory
KASS and VOS · Geometrical Foundations of Asymptotic Inference
KAUFMAN and ROUSSEEUW · Finding Groups in Data: An Introduction to Cluster Analysis
LAMPERTI · Probability: A Survey of the Mathematical Theory, *Second Edition*
LARSON · Introduction to Probability Theory and Statistical Inference, *Third Edition*
LESSLER and KALSBEEK · Nonsampling Error in Surveys
LINDVALL · Lectures on the Coupling Method
McFADDEN · Management of Data in Clinical Trials
MANTON, WOODBURY, and TOLLEY · Statistical Applications Using Fuzzy Sets
MARDIA · The Art of Statistical Science: A Tribute to G. S. Watson
MORGENTHALER and TUKEY · Configural Polysampling: A Route to Practical Robustness
MUIRHEAD · Aspects of Multivariate Statistical Theory
OLIVER and SMITH · Influence Diagrams, Belief Nets and Decision Analysis
*PARZEN · Modern Probability Theory and Its Applications
PRESS · Bayesian Statistics: Principles, Models, and Applications
PUKELSHEIM · Optimal Experimental Design
PURI and SEN · Nonparametric Methods in General Linear Models
PURI, VILAPLANA, and WERTZ · New Perspectives in Theoretical and Applied Statistics
RAO · Asymptotic Theory of Statistical Inference
RAO · Linear Statistical Inference and Its Applications, *Second Edition*
*RAO and SHANBHAG · Choquet-Deny Type Functional Equations with Applications to Stochastic Models
RENCHER · Methods of Multivariate Analysis
ROBERTSON, WRIGHT, and DYKSTRA · Order Restricted Statistical Inference
ROGERS and WILLIAMS · Diffusions, Markov Processes, and Martingales, Volume I: Foundations, *Second Edition;* Volume II: Îto Calculus
ROHATGI · An Introduction to Probability Theory and Mathematical Statistics
ROSS · Stochastic Processes
RUBINSTEIN · Simulation and the Monte Carlo Method
RUBINSTEIN and SHAPIRO · Discrete Event Systems: Sensitivity Analysis and Stochastic Optimization by the Score Function Method
RUZSA and SZEKELY · Algebraic Probability Theory
SCHEFFE · The Analysis of Variance
SEBER · Linear Regression Analysis
SEBER · Multivariate Observations
SEBER and WILD · Nonlinear Regression
SERFLING · Approximation Theorems of Mathematical Statistics
SHORACK and WELLNER · Empirical Processes with Applications to Statistics
SMALL and McLEISH · Hilbert Space Methods in Probability and Statistical Inference
STAPLETON · Linear Statistical Models
STAUDTE and SHEATHER · Robust Estimation and Testing
STOYANOV · Counterexamples in Probability
STYAN · The Collected Papers of T. W. Anderson: 1943–1985
TANAKA · Time Series Analysis: Nonstationary and Noninvertible Distribution Theory
THOMPSON and SEBER · Adaptive Sampling
WELSH · Aspects of Statistical Inference
WHITTAKER · Graphical Models in Applied Multivariate Statistics
YANG · The Construction Theory of Denumerable Markov Processes

*Now available in a lower priced paperback edition in the Wiley Classics Library.

Applied Probability and Statistics
   ABRAHAM and LEDOLTER · Statistical Methods for Forecasting
   AGRESTI · Analysis of Ordinal Categorical Data
   AGRESTI · Categorical Data Analysis
   AGRESTI · An Introduction to Categorical Data Analysis
   ANDERSON and LOYNES · The Teaching of Practical Statistics
   ANDERSON, AUQUIER, HAUCK, OAKES, VANDAELE, and WEISBERG ·
      Statistical Methods for Comparative Studies
   ARMITAGE and DAVID (editors) · Advances in Biometry
  *ARTHANARI and DODGE · Mathematical Programming in Statistics
   ASMUSSEN · Applied Probability and Queues
  *BAILEY · The Elements of Stochastic Processes with Applications to the Natural
      Sciences
   BARNETT and LEWIS · Outliers in Statistical Data, *Second Edition*
   BARTHOLOMEW, FORBES, and McLEAN · Statistical Techniques for Manpower
      Planning, *Second Edition*
   BATES and WATTS · Nonlinear Regression Analysis and Its Applications
   BECHHOFER, SANTNER, and GOLDSMAN · Design and Analysis of Experiments for
      Statistical Selection, Screening, and Multiple Comparisons
   BELSLEY · Conditioning Diagnostics: Collinearity and Weak Data in Regression
   BELSLEY, KUH, and WELSCH · Regression Diagnostics: Identifying Influential
      Data and Sources of Collinearity
   BERRY · Bayesian Analysis in Statistics and Econometrics: Essays in Honor of Arnold
      Zellner
   BERRY, CHALONER, and GEWEKE · Bayesian Analysis in Statistics and
      Econometrics: Essays in Honor of Arnold Zellner
   BHAT · Elements of Applied Stochastic Processes, *Second Edition*
   BHATTACHARYA and WAYMIRE · Stochastic Processes with Applications
   BIEMER, GROVES, LYBERG, MATHIOWETZ, and SUDMAN · Measurement
      Errors in Surveys
   BIRKES and DODGE · Alternative Methods of Regression
   BLOOMFIELD · Fourier Analysis of Time Series: An Introduction
   BOLLEN · Structural Equations with Latent Variables
   BOULEAU · Numerical Methods for Stochastic Processes
   BOX · R. A. Fisher, the Life of a Scientist
   BOX and DRAPER · Empirical Model-Building and Response Surfaces
   BOX and DRAPER · Evolutionary Operation: A Statistical Method for Process
      Improvement
   BOX, HUNTER, and HUNTER · Statistics for Experimenters: An Introduction to
      Design, Data Analysis, and Model Building
   BOX and LUCEÑO · Statistical Control by Monitoring and Feedback Adjustment
   BROWN and HOLLANDER · Statistics: A Biomedical Introduction
   BUCKLEW · Large Deviation Techniques in Decision, Simulation, and Estimation
   BUNKE and BUNKE · Nonlinear Regression, Functional Relations and Robust
      Methods: Statistical Methods of Model Building
   CHATTERJEE and HADI · Sensitivity Analysis in Linear Regression
   CHATTERJEE and PRICE · Regression Analysis by Example, *Second Edition*
   CLARKE and DISNEY · Probability and Random Processes: A First Course with
      Applications, *Second Edition*
   COCHRAN · Sampling Techniques, *Third Edition*
  *COCHRAN and COX · Experimental Designs, *Second Edition*
   CONOVER · Practical Nonparametric Statistics, *Second Edition*
   CONOVER and IMAN · Introduction to Modern Business Statistics

*Now available in a lower priced paperback edition in the Wiley Classics Library.

Applied Probability and Statistics (Continued)

    CORNELL · Experiments with Mixtures, Designs, Models, and the Analysis of Mixture Data, *Second Edition*

    COX · A Handbook of Introductory Statistical Methods

   *COX · Planning of Experiments

    COX, BINDER, CHINNAPPA, CHRISTIANSON, COLLEDGE, and KOTT · Business Survey Methods

    CRESSIE · Statistics for Spatial Data, *Revised Edition*

    DANIEL · Applications of Statistics to Industrial Experimentation

    DANIEL · Biostatistics: A Foundation for Analysis in the Health Sciences, *Sixth Edition*

    DAVID · Order Statistics, *Second Edition*

   *DEGROOT, FIENBERG, and KADANE · Statistics and the Law

   *DEMING · Sample Design in Business Research

    DETTE and STUDDEN · The Theory of Canonical Moments with Applications in Statistics, Probability, and Analysis

    DILLON and GOLDSTEIN · Multivariate Analysis: Methods and Applications

    DODGE and ROMIG · Sampling Inspection Tables, *Second Edition*

    DOWDY and WEARDEN · Statistics for Research, *Second Edition*

    DRAPER and SMITH · Applied Regression Analysis, *Second Edition*

    DUNN · Basic Statistics: A Primer for the Biomedical Sciences, *Second Edition*

    DUNN and CLARK · Applied Statistics: Analysis of Variance and Regression, *Second Edition*

    ELANDT-JOHNSON and JOHNSON · Survival Models and Data Analysis

    EVANS, PEACOCK, and HASTINGS · Statistical Distributions, *Second Edition*

    FISHER and VAN BELLE · Biostatistics: A Methodology for the Health Sciences

    FLEISS · The Design and Analysis of Clinical Experiments

    FLEISS · Statistical Methods for Rates and Proportions, *Second Edition*

    FLEMING and HARRINGTON · Counting Processes and Survival Analysis

    FLURY · Common Principal Components and Related Multivariate Models

    GALLANT · Nonlinear Statistical Models

    GLASSERMAN and YAO · Monotone Structure in Discrete-Event Systems

    GNANADESIKAN · Methods for Statistical Data Analysis of Multivariate Observations, *Second Edition*

    GREENWOOD and NIKULIN · A Guide to Chi-Squared Testing

    GROSS and HARRIS · Fundamentals of Queueing Theory, *Second Edition*

    GROVES · Survey Errors and Survey Costs

    GROVES, BIEMER, LYBERG, MASSEY, NICHOLLS, and WAKSBERG · Telephone Survey Methodology

    HAHN and MEEKER · Statistical Intervals: A Guide for Practitioners

    HAND · Discrimination and Classification

   *HANSEN, HURWITZ, and MADOW · Sample Survey Methods and Theory, Volume I: Methods and Applications

   *HANSEN, HURWITZ, and MADOW · Sample Survey Methods and Theory, Volume II: Theory

    HEIBERGER · Computation for the Analysis of Designed Experiments

    HELLER · MACSYMA for Statisticians

    HINKELMAN and KEMPTHORNE: · Design and Analysis of Experiments, Volume 1: Introduction to Experimental Design

    HOAGLIN, MOSTELLER, and TUKEY · Exploratory Approach to Analysis of Variance

    HOAGLIN, MOSTELLER, and TUKEY · Exploring Data Tables, Trends and Shapes

    HOAGLIN, MOSTELLER, and TUKEY · Understanding Robust and Exploratory Data Analysis

    HOCHBERG and TAMHANE · Multiple Comparison Procedures

    HOCKING · Methods and Applications of Linear Models: Regression and the Analysis of Variables

*Now available in a lower priced paperback edition in the Wiley Classics Library.

Applied Probability and Statistics (Continued)

HOEL · Elementary Statistics, *Fifth Edition*
HOGG and KLUGMAN · Loss Distributions
HOLLANDER and WOLFE · Nonparametric Statistical Methods
HOSMER and LEMESHOW · Applied Logistic Regression
HØYLAND and RAUSAND · System Reliability Theory: Models and Statistical Methods
HUBERTY · Applied Discriminant Analysis
IMAN and CONOVER · Modern Business Statistics
JACKSON · A User's Guide to Principle Components
JOHN · Statistical Methods in Engineering and Quality Assurance
JOHNSON · Multivariate Statistical Simulation
JOHNSON and BALAKRISHNAN · Advances in the Theory and Practice of Statistics: A Volume in Honor of Samuel Kotz
JOHNSON and KOTZ · Distributions in Statistics
    Continuous Multivariate Distributions
JOHNSON, KOTZ, and BALAKRISHNAN · Continuous Univariate Distributions, Volume 1, *Second Edition*
JOHNSON, KOTZ, and BALAKRISHNAN · Continuous Univariate Distributions, Volume 2, *Second Edition*
JOHNSON, KOTZ, and BALAKRISHNAN · Discrete Multivariate Distributions
JOHNSON, KOTZ, and KEMP · Univariate Discrete Distributions, *Second Edition*
JUDGE, GRIFFITHS, HILL, LÜTKEPOHL, and LEE · The Theory and Practice of Econometrics, *Second Edition*
JUDGE, HILL, GRIFFITHS, LÜTKEPOHL, and LEE · Introduction to the Theory and Practice of Econometrics, *Second Edition*
JUREČKOVÁ and SEN · Robust Statistical Procedures: Aymptotics and Interrelations
KADANE · Bayesian Methods and Ethics in a Clinical Trial Design
KADANE AND SCHUM · A Probabilistic Analysis of the Sacco and Vanzetti Evidence
KALBFLEISCH and PRENTICE · The Statistical Analysis of Failure Time Data
KASPRZYK, DUNCAN, KALTON, and SINGH · Panel Surveys
KISH · Statistical Design for Research
*KISH · Survey Sampling
LAD · Operational Subjective Statistical Methods: A Mathematical, Philosophical, and Historical Introduction
LANGE, RYAN, BILLARD, BRILLINGER, CONQUEST, and GREENHOUSE · Case Studies in Biometry
LAWLESS · Statistical Models and Methods for Lifetime Data
LE · Applied Survival Analysis
LEBART, MORINEAU., and WARWICK · Multivariate Descriptive Statistical Analysis: Correspondence Analysis and Related Techniques for Large Matrices
LEE · Statistical Methods for Survival Data Analysis, *Second Edition*
LePAGE and BILLARD · Exploring the Limits of Bootstrap
LEVY and LEMESHOW · Sampling of Populations: Methods and Applications
LINHART and ZUCCHINI · Model Selection
LITTLE and RUBIN · Statistical Analysis with Missing Data
LYBERG · Survey Measurement
MAGNUS and NEUDECKER · Matrix Differential Calculus with Applications in Statistics and Econometrics
MAINDONALD · Statistical Computation
MALLOWS · Design, Data, and Analysis by Some Friends of Cuthbert Daniel
MANN, SCHAFER, and SINGPURWALLA · Methods for Statistical Analysis of Reliability and Life Data
MASON, GUNST, and HESS · Statistical Design and Analysis of Experiments with Applications to Engineering and Science

*Now available in a lower priced paperback edition in the Wiley Classics Library.

Applied Probability and Statistics (Continued)

McLACHLAN and KRISHNAN · The EM Algorithm and Extensions
McLACHLAN · Discriminant Analysis and Statistical Pattern Recognition
McNEIL · Epidemiological Research Methods
MILLER · Survival Analysis
MONTGOMERY and MYERS · Response Surface Methodology: Process and Product in Optimization Using Designed Experiments
MONTGOMERY and PECK · Introduction to Linear Regression Analysis, *Second Edition*
NELSON · Accelerated Testing, Statistical Models, Test Plans, and Data Analyses
NELSON · Applied Life Data Analysis
OCHI · Applied Probability and Stochastic Processes in Engineering and Physical Sciences
OKABE, BOOTS, and SUGIHARA · Spatial Tesselations: Concepts and Applications of Voronoi Diagrams
OSBORNE · Finite Algorithms in Optimization and Data Analysis
PANKRATZ · Forecasting with Dynamic Regression Models
PANKRATZ · Forecasting with Univariate Box-Jenkins Models: Concepts and Cases
PIANTADOSI · Clinical Trials: A Methodologic Perspective
PORT · Theoretical Probability for Applications
PUTERMAN · Markov Decision Processes: Discrete Stochastic Dynamic Programming
RACHEV · Probability Metrics and the Stability of Stochastic Models
RÉNYI · A Diary on Information Theory
RIPLEY · Spatial Statistics
RIPLEY · Stochastic Simulation
ROSS · Introduction to Probability and Statistics for Engineers and Scientists
ROUSSEEUW and LEROY · Robust Regression and Outlier Detection
RUBIN · Multiple Imputation for Nonresponse in Surveys
RYAN · Modern Regression Methods
RYAN · Statistical Methods for Quality Improvement
SCHOTT · Matrix Analysis for Statistics
SCHUSS · Theory and Applications of Stochastic Differential Equations
SCOTT · Multivariate Density Estimation: Theory, Practice, and Visualization
*SEARLE · Linear Models
SEARLE · Linear Models for Unbalanced Data
SEARLE · Matrix Algebra Useful for Statistics
SEARLE, CASELLA, and McCULLOCH · Variance Components
SKINNER, HOLT, and SMITH · Analysis of Complex Surveys
STOYAN, KENDALL, and MECKE · Stochastic Geometry and Its Applications, *Second Edition*
STOYAN and STOYAN · Fractals, Random Shapes and Point Fields: Methods of Geometrical Statistics
THOMPSON · Empirical Model Building
THOMPSON · Sampling
TIERNEY · LISP-STAT: An Object-Oriented Environment for Statistical Computing and Dynamic Graphics
TIJMS · Stochastic Modeling and Analysis: A Computational Approach
TITTERINGTON, SMITH, and MAKOV · Statistical Analysis of Finite Mixture Distributions
UPTON and FINGLETON · Spatial Data Analysis by Example, Volume I: Point Pattern and Quantitative Data
UPTON and FINGLETON · Spatial Data Analysis by Example, Volume II: Categorical and Directional Data
VAN RIJCKEVORSEL and DE LEEUW · Component and Correspondence Analysis
WEISBERG · Applied Linear Regression, *Second Edition*

*Now available in a lower priced paperback edition in the Wiley Classics Library.

Applied Probability and Statistics (Continued)
WESTFALL and YOUNG · Resampling-Based Multiple Testing: Examples and Methods for *p*-Value Adjustment
WHITTLE · Optimization Over Time: Dynamic Programming and Stochastic Control, Volume I and Volume II
WHITTLE · Systems in Stochastic Equilibrium
WONNACOTT and WONNACOTT · Econometrics, *Second Edition*
WONNACOTT and WONNACOTT · Introductory Statistics, *Fifth Edition*
WONNACOTT and WONNACOTT · Introductory Statistics for Business and Economics, *Fourth Edition*
WOODING · Planning Pharmaceutical Clinical Trials: Basic Statistical Principles
WOOLSON · Statistical Methods for the Analysis of Biomedical Data
*ZELLNER · An Introduction to Bayesian Inference in Econometrics

Tracts on Probability and Statistics
BILLINGSLEY · Convergence of Probability Measures
KELLY · Reversability and Stochastic Networks
TOUTENBURG · Prior Information in Linear Models

*Now available in a lower priced paperback edition in the Wiley Classics Library.

Printed in the United States
1391200002BA/49